The Dynamic Greenhouse Effect and the Climate Averaging Paradox

Roy Clark

Ventura Photonics Monograph VPM 001
Ventura Photonics
Thousand Oaks
California

The Dynamic Greenhouse Effect and the Climate Averaging Paradox

Ventura Photonics Monograph VPM 001

Copyright © Roy Clark 2011

ISBN-13: 978-1466359185

ISBN-10: 1466359188

Ventura Photonics
Thousand Oaks
California

CONTENTS

PREFACE

The idea that variations in atmospheric carbon dioxide concentration could cause changes to the Earth's climate originated in the middle of the nineteenth century with the work of John Tyndall. Based on the limited knowledge at that time it was a reasonable explanation of the cause of an Ice Age. This idea became accepted as dogma over the years and was reinforced a century later by the discovery of changes in atmospheric carbon dioxide concentration from deep drilled ice core data. As the Earth warmed from an Ice Age, the atmospheric carbon dioxide concentration increased from about 200 to 280 parts per million. However, we now know that this was caused by the decrease in the ocean solubility of carbon dioxide as the oceans warmed up. The sun had to heat the oceans first. Climate change is caused by small fluctuations in the solar constant that accumulate as heat over long periods in the Earth's oceans. The solar constant, the average solar flux reaching the Earth is 1365 $W.m^{-2}$. This increases by about 1 $W.m^{-2}$ when the sun is active and the sunspot index is near 100. It changes by a similar amount as the planets, particularly Jupiter change the ellipticity of the Earth's orbit over a nominal 100,000 year Ice Age cycle. Simple heat capacity calculations show that it only requires an increase of 0.4 $W.m^{-2}$ in the solar flux at the surface to warm the Earth out of an Ice Age, although it requires 10,000 years for the ice to melt.

Unfortunately, modern academic research is judged, not by its scientific validity, but by the magnitude of the research funds obtained and the number of papers published. Starting in the 1960's, exaggerated claims of carbon dioxide induced global warming were used as a funding vehicle for a wide range of research topics. First, carbon dioxide induced global warming was used as an excuse by infra-red spectroscopists and astronomers to fund research on radiative transfer in planetary atmospheres. This was an exercise in applied mathematics not physics. The dynamic energy transfer processes that determine the Earth's surface temperature were ignored. Simplistic equilibrium assumptions were made that allowed calculations to be performed using the limited computer capabilities available at the time. This was before moon landings and weather satellites. These researchers were at least honest and published their assumptions, including a zero heat capacity for the Earth's surface. The 'equilibrium average surface temperature' calculated by these models is not even a physically measurable climate variable. However, the mathematical theory had some elegance and allowed some limited predictions to be made about the effects of carbon dioxide on climate. The fact that these predictions were invalid was ignored. No-one ever bothered to check the models against physical reality. However, the idea that the models were calculating some form of surface temperature was still accepted. The real interest was in the IR emission from the top of the planetary atmosphere, not the surface energy transfer.

Starting in the early 1980's, funding for atomic energy research began to dwindle and some researchers who were familiar with carbon 13 isotope ratios began to use carbon dioxide induced global warming as an excuse for research funding. The limitations of the existing radiative transfer based climate models were ignored and the great global warming fraud started to take off. The fundamental error was the substitution of the meteorological surface air temperature record for the real surface temperature. This is the temperature used in weather forecasting. It is the air temperature that is measured in a vented enclosure placed for convenience at eye level above the ground. It has nothing to do with the real surface temperature below the enclosure. The dominant source of long term variation in the air temperature record is not the local surface temperature. It is usually the change in ocean surface temperatures in the region of origin of the prevailing weather systems reaching the weather station. There are natural ocean surface temperature cycles with a period of about 60 years. 'Global warming' just happened to coincide with the 30 year warming cycle of the Pacific Decadal Oscillation. In the US, this was just a repeat of the dust bowl temperature peak in the 1930's. Furthermore, urban growth around some of the weather stations introduced additional local heating measurement biases that added to the real weather temperatures. These are now known as urban heat island effects. Later, since the same climate 'scientists' controlled both the model predictions and the climate record, 'adjustments' were made to the weather data to exaggerate the climate change.

The climate models were empirically hard wired using radiative forcing 'constants' to produce global warming whether it existed or not. The increase in the 'average global meteorological surface air temperature' was used to 'calibrate' the increase in 'surface temperature' produced by carbon dioxide. It was simply assumed that carbon dioxide had to be the cause of the observed temperature change. The fictional concept of an average 'climate sensitivity' to carbon dioxide was introduced. This was extended to other infra red active species in the atmosphere using the pseudoscience of 'radiative forcing constants'. 'Greenhouse gases' produced warming and 'aerosols' produced cooling. The two could be empirically balanced to produce any desired result. 'Volcanic aerosols' were used for fine tuning. All of the large scale climate models that use this concept of radiative forcing are based on pure pseudoscience. Belief in computer based climate astrology has replaced climate science.

The science no longer mattered all the while the money kept on rolling in. Over a trillion dollars has been consumed by this climate fraud. Government agencies were funded to study global warming. NASA wanted more satellites and took whatever climate money it could get. The satellites of course had to measure 'global warming'. No other interpretation of the data was allowed. NOAA is still lying about 'extreme climate events' caused by global warming. DOE is still funding carbon sequestration. The peer review process in climate science collapsed and was replaced by flagrant cronyism. Those that had benefited from global warming funds actively opposed any change in global warming dogma that would threaten their funding. The leading climate 'scientists' became trapped in a web of their own lies. Many other researchers jumped on the global warming gravy train to 'save the planet from global warming hell'. Leading scientific societies, including the Royal Society of London were fooled into supporting climate astrology. This has led to the alternative energy fiasco and carbon taxes that are a recipe for economic disaster. Even banks and insurance companies have made large investments based on fraudulent global warming claims.

While the global warming fraud has long been known and tolerated in scientific circles because of the money involved, it has been hidden from wider public view by a very effective propaganda campaign. This has been dominated by the UN IPCC, the United Nations Intergovernmental Panel on Climate Change. The fraudulent claims that led to the Kyoto Treaty were first revealed to a larger audience in November 2009 when the so called 'Climategate' archive of e-mails and other electronic files was published on the web. This exposed a systematic pattern of egregious fraud that is still under investigation. 'Amazongate', 'Glaciergate' and other climate frauds have now been added to the list. The stranglehold of global warming on climate research is gradually diminishing.

However, global warming dogma has been successfully perpetuated for over forty years. A whole generation of researchers has grown up surrounded by climate fraud and the culture of lying for money. In order to return to physical reality it is necessary to go back to first principles and establish how the Earth's climate really works. This requires a fundamental change in the way we think about climate energy transfer. There is no such thing as climate equilibrium on any time scale. The energy transfer processes that determine the Earth's surface temperature are dynamic, not static. The simple mathematical elegance of a few fraudulent flux equations has to be replaced with the more complex description of a dynamic 'greenhouse effect'. There are six energy transfer processes that interact dynamically with four thermal reservoirs to determine the Earth's climate. Temperature is defined in terms of the heat content of these thermal reservoirs and this is always changing.

The dynamic energy transfer also leads to a climate averaging paradox. The Earth's climate is the result of a long term accumulation of short term weather changes. There is no 'equilibrium average' shortcut, so climate change has to be calculated the way it is measured, as a long term average of short term changes. The maximum time step used in the analyses presented in this monograph is one hour. Even the ocean heating produced by 450 years of sunspots was calculated as a series of half hour averages.

It is hoped that the research results presented here will help to guide climate science back to its foundations in meteorology. The starting point is the basic laws of physics, particularly the First and Second Laws of Thermodynamics applied to the measured surface energy transfer. The Earth's climate is an open cycle heat engine, not an infra-red spectrometer. There is a basic need for more detailed and widespread measurements of the dynamic surface energy transfer processes. The primary function of climate science is to explain existing climate observations. Only when this is successful can any attempt at future climate prediction be made using properly validated climate models.

The lessons learned from the global warming fraud also have profound implications in other areas such as alternative energy and environmental science. The extreme exaggeration of the problem and use of invalid computer models is not limited to climate astrology. How much pseudoscience has already found its way into alternative energy and environmental science? Where does the science end and 'green religion' begin? It is time for a thorough investigation into scientific fraud in all areas of environmental science and alternative energy. Global warming is not an isolated case of fraud.

1.0 INTRODUCTION

The Earth's surface is warmer than it should be, based on the average amount of solar radiation that it absorbs. However, in order to explain this correctly, we need to stop thinking in terms of averages and climate equilibrium states. The energy transfer processes that determine the Earth's surface temperature are not static, but dynamic. The surface is always heating and cooling on both a daily and a seasonal time scale. The sun heats the surface during the day and the temperature depends on the balance between the heating and cooling fluxes. There is no such thing as 'climate equilibrium'. Because of the heat storage by the surface, there is a delay between the peak solar flux and the peak temperature. This is clear evidence that there is no thermal equilibrium. The surface cools through a combination of convection, evaporation and emission of long wave infrared (LWIR) radiation. The IR active gases in the atmosphere, mainly water, H_2O and carbon dioxide, CO_2, also emit thermal LWIR radiation. The downward LWIR flux from the lower troposphere slows the cooling of the surface through a process of radiation exchange. However, this is only one part of a complex dynamic energy transfer process and the dominant cooling flux from the surface is usually moist convection as the surface is warmed by the sun during the day. The downward LWIR flux from the atmosphere does not determine or control the surface temperature and there is no equilibrium on any time scale.

The role of LWIR emission by the atmosphere was described by Joseph Fourier in 1827 and speculation that changes in the atmospheric concentration of CO_2 could cause the Ice Age cycle dates back to John Tyndall in about 1860.[1,2] However, we now know that climate changes are not caused by CO_2, but by small changes in the solar energy coupled into the oceans and by changes in ocean circulation.[3] Conclusive evidence for the solar origin of Ice Ages has only become available in recent years. This is derived from the analysis of ice cores and ocean sediments and from satellite observations of climate data and solar radiation. Unfortunately, the empirical speculation of CO_2 induced climate change - 'global warming' - has become accepted as dogma. Economic and political factors have made this dogma rather difficult to change. A large body of empirical pseudoscience has been produced using invalid climate simulation techniques based on the concept of 'radiative forcing'.[4] Every imaginable form of disaster that can even remotely be linked to climate change has been attributed to 'global warming'. None of this has any basis in physical reality. The peer review process in climate science has collapsed and numerous pseudoscientific papers on climate change have been published in 'respected' scientific journals. Such papers should never have seen the light of day. These fraudulent climate models have given the illusion of scientific respectability to the publications of the United Nations Intergovernmental Panel on Climate Change, IPCC.[5] However, it is now clear that the IPCC is a political body and the IPCC publications are nothing more than propaganda.[6] Radiative forcing assumes that long term averages of dynamic climate variables such as surface temperature somehow describe an 'equilibrium state' that can be perturbed by small changes in atmospheric CO_2 concentration. This is nothing more than a mathematical abstraction that was needed to simplify climate analysis in the 1960s. In reality, the daily changes in surface flux and surface temperature are so large that the climate effects of a 100 ppm increase in atmospheric CO_2 concentration are too small to measure. The whole issue of CO_2 induced global warming disappears once the greenhouse effect is considered as a dynamic, time dependent process.

The greenhouse effect is usually described in terms of a very simplistic equilibrium energy balance. The Earth's 'average surface temperature' is 288 K (15 C). The average 'effective atmospheric emission temperature' is 255 K (-18 C). This is the blackbody temperature that produces the long wave infrared (LWIR) emission needed balance the average solar flux absorbed by the surface. The temperature difference, 288 − 255 = 33 K is the surface temperature increase caused by 'greenhouse gas absorption'. It is then empirically assumed that small increases in downward atmospheric LWIR flux from an increase in 'greenhouse gas concentration' will cause an increase in 'surface temperature'.[7] However, there is a fundamental flaw in this argument. In order for heat to flow, there has to be a thermal gradient, or temperature difference. The surface has to be warmer than the air above. This is required by the Second Law of Thermodynamics. When the sun heats the surface during the day, the dominant cooling process is usually moist convection, not the emission of LWIR radiation. There can be no surface equilibrium on any time scale. As the warm air rises through the atmosphere it expands and cools. At the standard lapse rate of -6.5 K.km^{-1}, the air has to rise to an altitude of 5 km to cool by 33 K. This is just the 'average emission height' for atmospheric LWIR emission. The transfer of energy from the surface is controlled dynamically by moist convection, not by any kind of 'average equilibrium flux' or 'greenhouse gas absorption'. This surface convection 'resets' the local lapse rate and therefore the local atmospheric temperature and water vapor profiles on a daily basis. The atmosphere is warmed by a daily convective 'pulse' and cools through LWIR emission to space produced by the narrowing of the individual molecular lines with altitude. The convective heating and the LWIR cooling of the atmosphere are independent energy transfer processes. One does not control the other. The atmosphere continues to cool until it receives its next convective heating pulse. There is a dynamic energy balance and any change in this thermal balance produces a change in the height of the tropopause. This daily convective pulse also determines the downward LWIR emission from the atmosphere to the surface. Most of this flux originates from a 'thermal blanket' within the first kilometer layer above the surface.

The underlying assumption of radiative forcing, that long term averages of surface temperature are in some kind of equilibrium with the long wave infrared (LWIR) emission to space has no basis in physical reality. The tail does not wag the dog. There are six distinct physical processes that interact dynamically to determine the surface temperature and the LWIR radiation to space. These are:

1) Energy transfer at the air-land interface
2) Energy transfer at the air-ocean interface
3) Direct LWIR emission to space from the surface
4) Energy transfer between the lower troposphere and the surface
5) Upward convective transport to the middle and upper troposphere
6) LWIR emission to space.

However, before these six processes are considered in detail, a brief history of the greenhouse effect and the basic technical background needed to understand the energy transfer processes will be presented.

2.0 HISTORICAL BACKGROUND

In his discussion of the temperature of the Earth in 1827, Joseph Fourier described experiments in which air was heated by sunlight in a black walled container sealed with glass windows.[1] This is the origin of the term 'greenhouse effect', although the atmosphere does not trap heat like a greenhouse. Here, Fourier was in fact describing the conversion of sunlight to heat. He also understood convection and dynamic surface heating, both daily and seasonal. These effects were clearly described by Fourier in his paper. In 1837, Louis Agassiz proposed his theory of glaciation, that there had been an Ice Age in the past.[8] This was based on his observations of glaciers in the Alps. It took about thirty years of rather acrimonious debate before the idea of an Ice Age became accepted.[9] Prevailing scientific and religious dogma still required a 'Great Flood'. We now know that there was not just one Ice Age, but many, extending back through the geological past.[10] Starting in 1859, John Tyndall began his studies of the infra red absorption of gases and correctly identified water vapor, followed by carbon dioxide as the most important IR absorbing gases in the atmosphere.[2] He was also interested in the study of glaciers and accepted the Ice Age glaciation theories of Louis Agassiz. This led him to propose that changes in CO_2 concentration might be responsible for climate change. This idea was resurrected by Svante Arrhenius in 1896.[11] The first quantitative evidence for an increase in atmospheric CO_2 concentration was provided by Guy Stewart Callendar in 1938.[12] The relationship between CO_2 in the atmosphere and its intake by the oceans was first quantified by Roger Revelle and Hans Seuss in 1957.[13]

Starting in 1958, Charles David Keeling began to record the atmospheric CO_2 concentration at a weather observatory on the peak of Mauna Loa, Hawaii on a continuous, long term basis. This work continues today. The so called Keeling curve which shows the measured increase in atmospheric CO_2 concentration over time is shown in Figure 2-1.[14] The curve is correct and a tribute to the tenacity of the researchers that have maintained the record since 1958. However, its empirical use to justify global warming is one of the most egregious and expensive scientific frauds ever perpetrated. No quantitative energy transfer relationship has ever been established between the measured increase in CO_2 concentration and climate change. This is because none exists.

In order to connect the increase in CO_2 concentration to a change in surface temperature, a theoretical construct known as radiative forcing has been adopted. This was introduced into climate simulation in its present form by Manabe and Wetherald in 1967.[15]   It is assumed, without justification or validation, that long term averages of transient, non-equilibrium variables can be analyzed as a system that is in equilibrium. The upward and downward fluxes at some rather ill defined boundary such as an 'average tropopause' are assumed to be equal and equivalent. A change in CO_2 concentration is introduced to 'perturb' this 'equilibrium' and the change in flux is used to calculate a new 'equilibrium surface temperature'. This calculated 'equilibrium surface temperature' produced by such models has no physical meaning. It really assumes that the sun is shining all the time and that the unperturbed surface is always receiving and emitting a flux of 390 $W.m^{-2}$. The heat capacity of the surface is set to zero. Small, 1 to 4 $W.m^{-2}$ changes in flux in a stratospheric layer of air at 217 K and 0.22 atm. are assumed to be capable of warming a surface at 288 K through 11 km of warmer, higher density air. This requires a flagrant violation of the Second Law of Thermodynamics. The cooling of the surface by convection, the conversion of IR radiation into other forms of energy and the heat capacity of the surface are also conveniently ignored. Further details on the history of global warming can be found in Imbrie and Imbrie, 1979, Weart, 1997 and Ramanathan, 1998.[9,16,17]

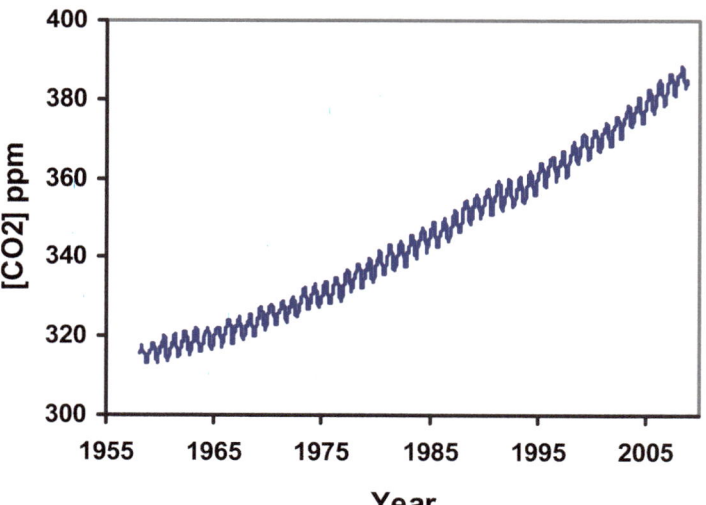

Figure 2-1: The Keeling curve: the increase in atmospheric CO_2 concentration from 1958.

However, the idea that an increase in CO_2 concentration must lead to an increase in surface temperature was initially accepted almost without question. Careful analysis of the long term meteorological surface air temperature (MSAT) record in the 1980s revealed a small increasing trend that was immediately correlated by empirical speculation to the rise in CO_2 concentration.[18] The dependence of the MSAT on weather patterns, ocean surface temperatures, solar illumination and surface absorption was ignored and empirical correlation is not proof. The underlying physics of the surface energy transfer was completely neglected. Two unrelated plots of meteorological temperature and CO_2 concentration were overlaid and made to coincide to produce the so called 'hockey stick' graph. Exaggerated concern over ozone depletion led to the inclusion of other greenhouse gases into the radiative forcing models. An elaborate set of radiative forcing constants that empirically related small changes in IR emission to surface temperatures was constructed.[4] This was 'calibrated' using the change in the MSAT. A mysterious 'water vapor feedback' was invoked to explain any discrepancies between the blackbody emission temperature and the surface temperature 'calibration'. This 'global warming' has no relationship to the true ground surface temperature that is needed to calculate the IR surface flux and no demonstrated causal relationship to the change in CO_2 concentration. This whole approach is pseudoscience. The same technique is used in astrology. While the positions of the planets can be calculated quite accurately, they have no relationship to human behavior. The only 'proof' ever provided for radiative forcing is that the results from one invalid computer model can be made to agree with those from another. No experimental verification is apparently required, nor can any direct measurement of 'equilibrium surface temperature' be performed.

Global warming conveniently ignores a similar temperature increase that occurred in the U. S. during the dust bowl droughts in the 1930s before there was any significant increase in atmospheric CO_2 concentration. The US MSAT anomaly record is shown in Figure 2-2.[19,20] The anomaly is just the temperature change with the average removed. The increases in black body surface temperature derived from the increase in downward CO_2 LWIR flux and from the 'radiative forcing constant' are also shown. Clearly, there is no relationship between the MSAT record and the presumed increase in 'equilibrium surface temperature' from CO_2 derived directly from the LWIR flux or from the forcing constant. The increase in blackbody 'equilibrium' temperature derived from the calculated increase in the downward flux is 0.3 C, based on an increase in atmospheric CO_2 concentration from 280 to 380 ppm. This has real physical meaning as the increase in temperature needed to increase the LWIR flux of a blackbody at 288 K (15 C) by 1.7 $W.m^{-2}$. The higher temperature increase derived from the forcing constant is purely empirical and assumes a mysterious water vapor 'feedback' effect. This temperature increase has no basis in physical reality. It should also be noted that the GISS climate record available from Reference 19 has been periodically 'adjusted' to reduce the dust bowl peak.[21] The climate record has been 'rigged' to support the 'predictions' of the climate astrologers.

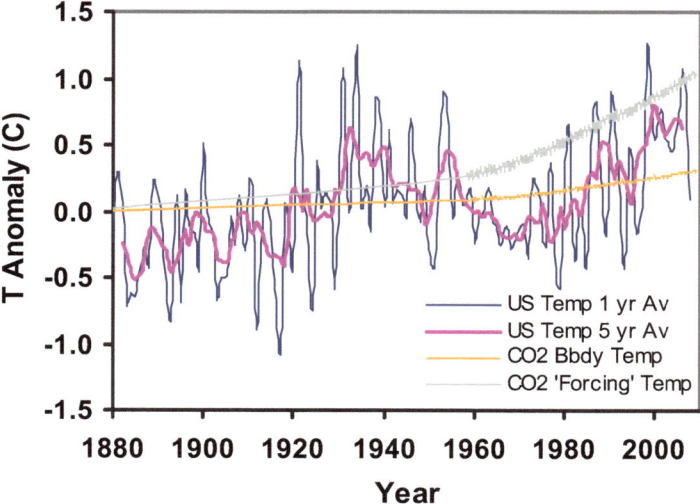

Figure 2-2: The US MSAT anomaly, 1 and 5 year averages from 1880. The temperature increases derived from the increase in downward LWIR CO_2 flux and from the corresponding forcing constant are also shown.

3.0 TECHNICAL BACKGROUND

In order to understand the dynamic energy transfer processes that determine the Earth's surface temperature, it is necessary to have some background in thermodynamics, fluid mechanics, black body radiation, atmospheric attenuation, the spectral properties of H_2O and CO_2, the lapse rate, the optical properties of the oceans and surface heat transfer. The basic concepts of these topics are presented below.

3.1 Thermodynamics and Fluid Mechanics

The energy transfer processes that determine the Earth's climate are complex and the detailed weather patterns that result can be chaotic. The basic physical principles needed to describe climate energy transfer were developed in the nineteenth century. However, it is only since electronic computers became available starting in the in 1950s that global scale weather prediction and climate simulation have become feasible. Thermodynamics began as theoretical attempts to explain the efficiencies, or rather the inefficiencies of early steam engines, starting with the work of Sadi Carnot in the 1820s. The basic principles of heat engines, described in terms of working fluids without getting into the details of atoms and molecules are governed by the First and Second Laws of Thermodynamics.

The First Law is simply the conservation of energy, which we take for granted today without much thought. However, conservation of energy still allows energy to be converted from one form to another. Sunlight is converted into heat and heat released into the atmosphere produces air pressure changes, convection currents and wind. Furthermore, conservation of energy does not imply radiative equilibrium or conservation of radiative flux. Under full summer sun conditions, a solar flux of 1 $kW.m^{-2}$ will result in a blackbody equilibrium temperature of 91.4 C. We are fortunate that this condition is never reached.

The Second Law in its simplest form states that heat cannot flow from a cooler to warmer body. If we are sitting in a coffee shop with two similar cups of coffee in front of us, both will continue to cool at similar rates while we are talking. One cup will not suddenly heat up on its own and cause the other one to become ice cold. The effects of the Second Law may be quantified using the concept of entropy, but for our purposes that is not required. However, we will need to use the Second Law to understand radiative transfer and Kirchoff exchange energies. This is considered below in Section 3.2 along with black body radiation. Another important point related to the Second Law is that heat transfer requires a thermal gradient. The surface can only transfer heat to the atmosphere when it is warmer than the air above.

A link between the trade winds and the rotation of the Earth was first proposed by George Hadley in 1735. In 1835, Gaspard Coriolis introduced the rotational force that now bears his name. The correct explanation for atmospheric circulation in terms of angular instead of linear momentum was published by William Ferrell in 1856.[22] As the warm air rises in the tropics, it does not have a sufficient angular velocity to keep up with the rotation of the Earth underneath. The Earth rotates towards the sun, from west to east and as we move into the rising air mass near the equator, we perceive a wind in the opposite direction. As the air rises, it cools and water condenses with cloud formation. The cool, dry air descends, warms up and creates the desert regions in the 30° latitude bands. The air is now rotating faster than the Earth and sets up the easterly trade winds. This circulation pattern is known as the Hadley cell. The trade winds couple to the ocean surface and set up the large scale ocean circulation gyres that are found in all of the world's oceans. Carl-Gustaf Rossby first explained large scale atmospheric motion using fluid dynamics in 1940 and was influential in the development of the first computer based weather predictions by John von Neuman starting in the late 1940's.[23] The effect of the Earth's rotation on wind induced ocean currents was first explained by Vagn Walfrid Ekman in 1902. This work was expanded by Johan Sandström, Harald Sverdrup, Henry Stommel and Walter Munk to a full description of both the wind driven surface and deep thermohaline ocean circulation.[24] The basic atmospheric circulation and wind patterns are illustrated in Figure 3-1 and the resulting ocean circulation is shown in Figure 3-2.

The warm ocean waters produced in the tropics are circulated by large scale ocean gyres that are driven by the trade winds. The warm water accumulates into large surface pools in the western tropical regions of the Atlantic and Pacific oceans. Part of this circulation is directed to higher latitudes by surface currents such as the Gulf Stream and the Kuroshio (Japan) current. When sea ice freezes at the poles, a cold high density brine is also produced that sinks and contributes to the thermohaline circulation. This is a deep ocean counterflow to the wind driven surface currents that will not be considered in detail here. The pattern of ocean surface temperatures produced by the wind driven surface currents sets the Earth's basic weather patterns through evaporation and atmospheric heating. However, the winds and currents form a complex interactive system that can exhibit both oscillatory and chaotic behavior. Ocean surface temperature differences and the associated air pressure gradients are used to define long term climate oscillations including the El Nino

Southern Oscillation, ENSO, the Pacific Decadal Oscillation, PDO and the Atlantic Multidecadal Oscillation, AMO.[25]

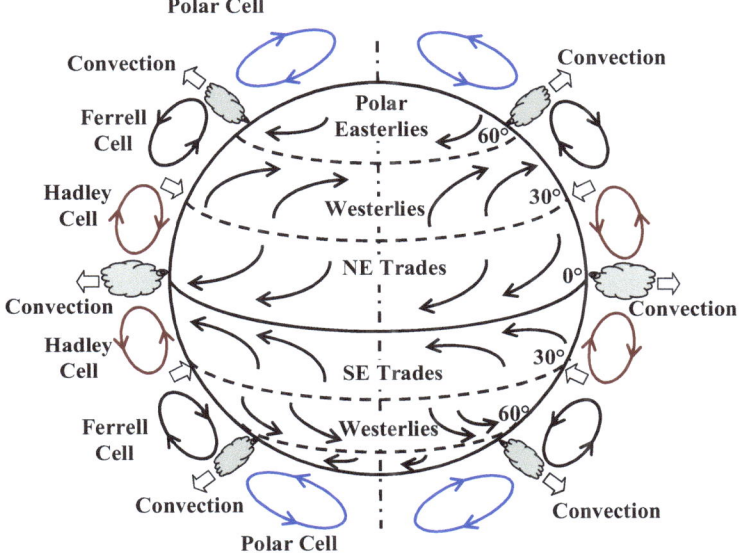

Figure 3-1: Atmospheric circulation and wind patterns (simplified schematic).

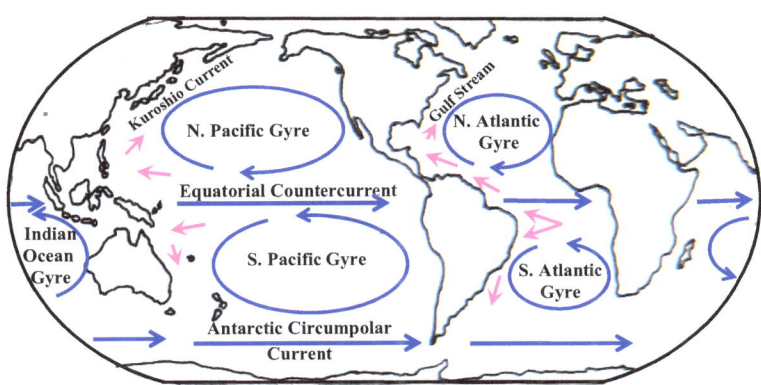

Figure 3-2: Wind driven ocean circulation (simplified schematic).

3.2 Black Body Radiation and the Kirchoff Exchange Energy

When we turn on the heating element of an electric cooker, it begins to get hot and radiate heat. We can feel the heat by holding our hand over the element without touching it. As the burner heats up it begins to glow dull red and at full heat it is bright red. Blackbody radiation temperatures are usually given in Kelvin, the absolute temperature scale obtained by adding 273 (more accurately 273.15) to the Celcius scale. When a heating element first starts to appear red, it is at a temperature of about 1000 K. At higher temperatures the blackbody emission color begins to turn from red to white and then to a bluish white. A tungsten filament lamp operates at a temperature of ~3000 K. A high intensity arc lamp operates at ~6000 K. The sun has a blackbody temperature of ~5800 K. The total amount of heat radiated by a blackbody increases rapidly with temperature. The total intensity relationship is governed by Stefan's Law, which states that the blackbody emission increases as the fourth power of the absolute temperature. This is shown in Figure 3-3. The peak intensity also shifts to shorter wavelengths as the temperature increases. The wavelength distribution is given by Planck's law. The shift of the peak emission wavelength is given by Wein's Law. This is illustrated in Figure 3-4.

The emission from a non-black body at a given wavelength and temperature is usually described in terms of an emission coefficient, ε. This is simply the ratio of the emission compared to a blackbody at the same wavelength. The absorption coefficient, α, at the same temperature and wavelength is equal to the emission coefficient. When a body is in equilibrium at a given temperature, it is absorbing and emitting equal amounts of radiation. This is one way of stating Kirchoff's Exchange Law. When two bodies are at the same temperature, they exchange equal amounts of thermal energy with each other. This follows from the Second Law of Thermodynamics. If there is no thermal gradient between the two bodies, then the same amount of heat must flow in both directions between them. If there is a difference in temperature between the two bodies, heat will flow from the warmer to the cooler body. The heat flux is the net difference in the exchange energy between the two bodies. This is a very important concept in surface energy transfer. The downward LWIR flux from the IR active species in the atmosphere balances most of the upward LWIR flux from the surface. This slows the rate of cooling of the surface. However, the surface air temperature is normally set during the day by the solar heating of the surface and the resulting moist convection. This is discussed in detail in Section 4.1 below.

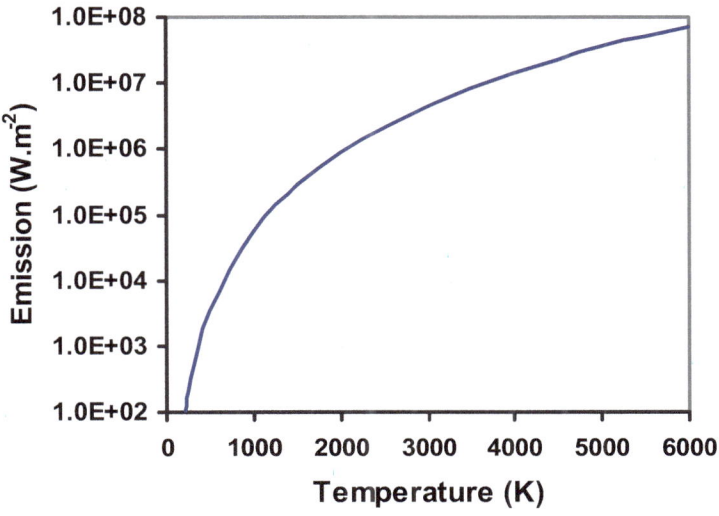

Figure 3-3: The increase in blackbody radiation with temperature.

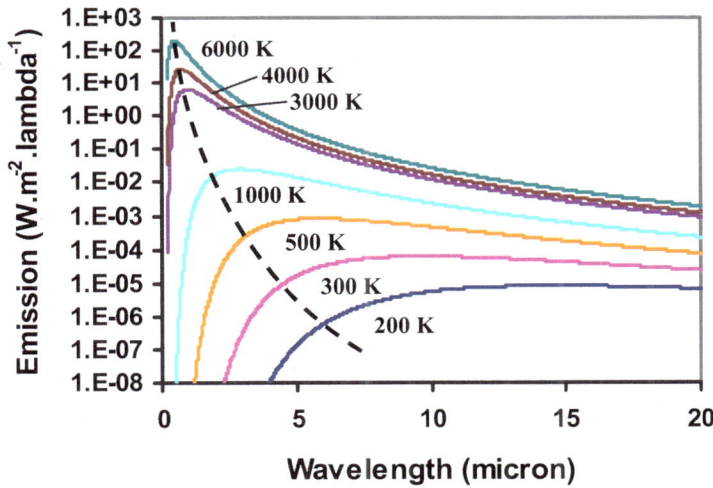

Figure 3-4: The wavelength dependence of blackbody radiation.

3.3 Atmospheric Attenuation

The solar constant, the average intensity of the solar flux at the top of the atmosphere is 1365 W.m^{-2}. The solar radiation reaching the Earth's surface at normal incidence is approximately 1000 W.m^{-2}. Below 200 nm, the UV radiation is absorbed by molecular oxygen, O_2. Some of the O_2 is dissociated by the UV radiation and the resulting atomic oxygen reacts with more O_2 to form ozone, O_3 in the upper atmosphere. This blocks the UV in the 200 to 300 nm region. Rayleigh scattering, the elastic scattering of light by air molecules reduces the solar intensity in the blue and near UV regions. Rayleigh scattering varies as the inverse fourth power of the wavelength. This produces the characteristic blue color of the sky. In addition, Rayleigh scattering, when viewed at 90° to the incident light is strongly linearly polarized. However, this polarization is not easily observed by the unaided human eye and plays a minor role in climate energy transfer. The atmosphere transmits most of the visible spectrum from the sun, which peaks in the green near 550 nm. In the near IR (NIR) region, the sunlight is attenuated by the vibrational overtones of the water spectrum. The solar spectrum at the top of the atmosphere and at the surface for an atmospheric optical depth of 1.5 (48.2° solar zenith angle) are shown in Figure 3-5.[26]

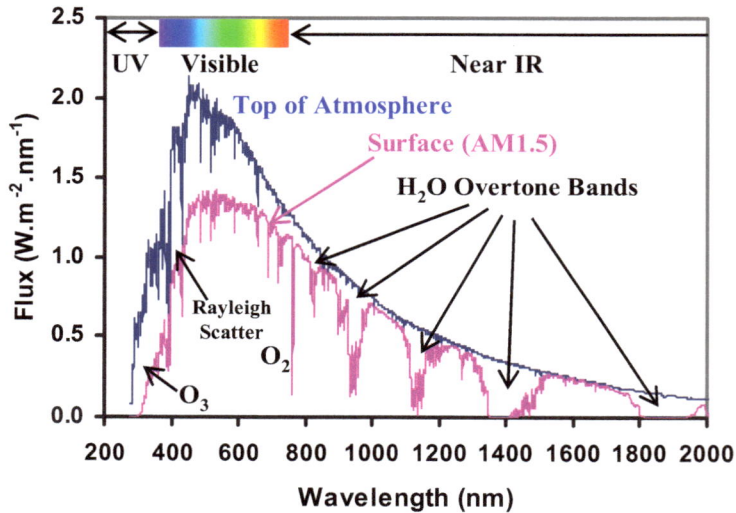

Figure 3-5: Atmospheric attenuation for top of the atmosphere, AM0 and for an atmospheric optical depth of 1.5, AM1.5.

3.4 The IR spectra of H_2O and CO_2 in the atmosphere

The atmospheric spectra observed in the LWIR region are associated with the absorption and emission of photons between molecular vibration-rotation states. These may be visualized using a simple rotating mechanical spring model in which the atoms of the molecule are connected by vibrating springs with rotations about the principal moments of inertia. A simple mathematical description of the vibrations is obtained by using molecular symmetry to define the normal modes of vibration. These are illustrated in Figure 3-6 for H_2O and CO_2. Both the rotation and vibration energies are quantized, or limited to discrete values. Quantum mechanical selection rules restrict the number of allowed transitions between the vibration-rotation states. Usually the quantum number change is limited to values of 0 and ±1. In simple cases, bands of lines may be observed corresponding to these changes in quantum number. Lines with quantum number changes $\Delta J = -1, 0, +1$ are designated as P, Q and R branches. These may be seen in the 15 μm CO_2 band. The spectroscopic notation predates the quantum mechanical explanation of the lines.

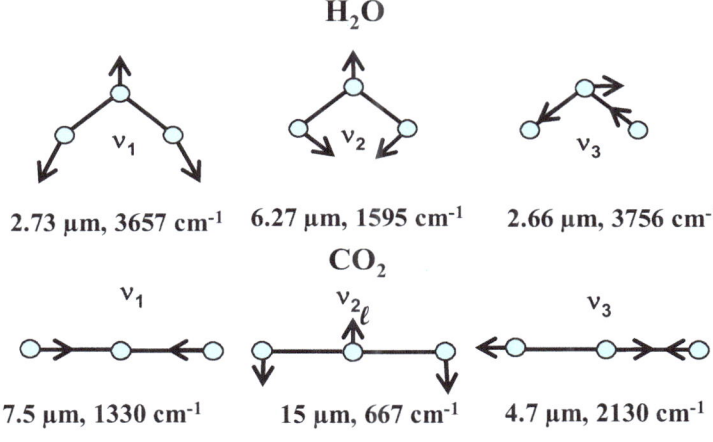

Figure 3-6: The normal modes of vibration of CO_2 and H_2O.

The principal absorption bands of interest in atmospheric radiative transfer are the pure rotation spectrum and the lowest vibration fundamental transition, v_2 of H_2O, and the lowest vibration fundamental transition, v_2 of CO_2. Each of these bands contains a large number of lines that are transitions between the rotational energy levels of the molecule in the upper and lower state. The pure rotation spectrum of H_2O occurs mainly below 550 cm^{-1}. The principal CO_2 lines occur between 550 and 800 cm^{-1}. The LWIR window occupies the region between 800 and 1200 cm^{-1}. The H_2O v_2 band occurs between 1200 and 2000 cm^{-1}. A plot of the linestrengths at 296K vs. position is shown in Figure 3-7. The actual line profile depends on the temperature, the pressure, the concentration and the path length, so there are several steps needed to calculate the line profile from the linestrength. The spectral data needed to calculate the high resolution IR spectra of most of the IR active molecules found in the atmosphere are available from the HITRAN data base.[27] The details of the calculations used to generate the plots used here are given in Reference 3. The spectral lines from other minor greenhouse gases such as ozone and methane are not included in Figure 3-7. Since it is demonstrated here that a 100 ppm increase in atmospheric CO_2 concentration can have no effect on the Earth's climate, increases in the atmospheric concentration of other minor 'greenhouse gases' will also have no effect on climate. There is no need to consider them in this analysis.

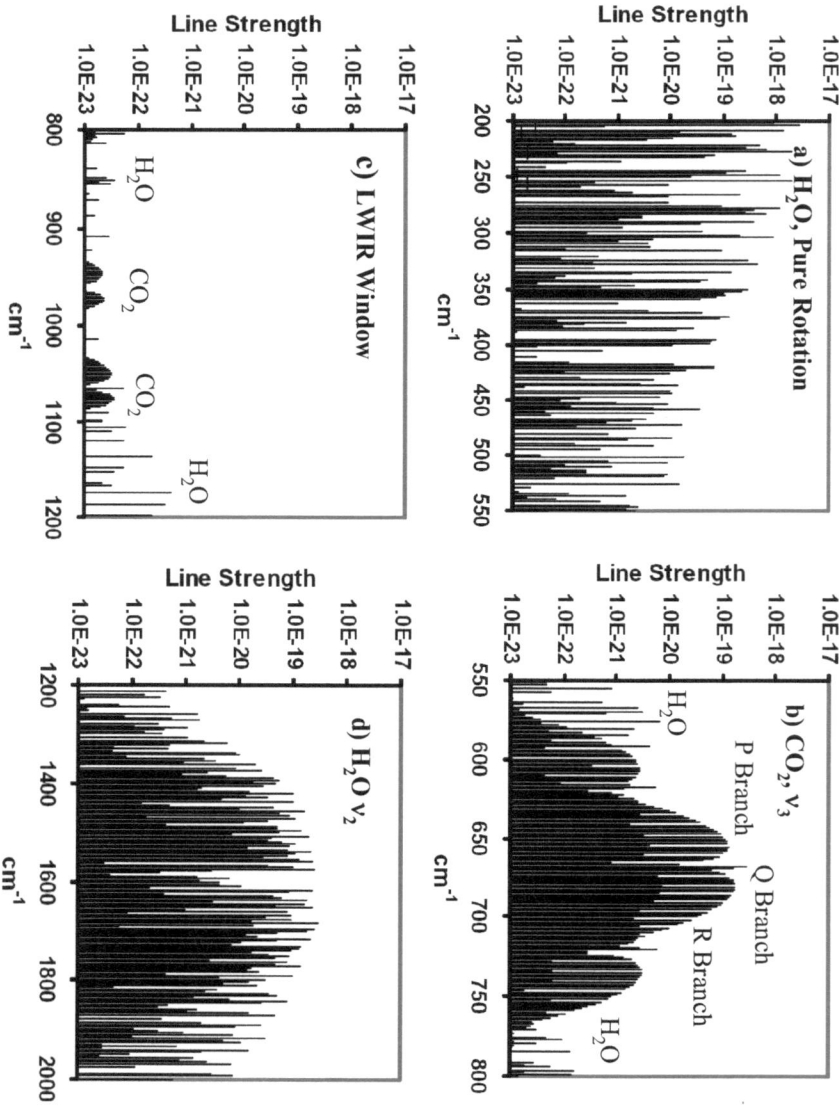

Figure 3-7: Line strengths of $^1H^{16}O_2$ and $^{12}C^{16}O_2$ from 200 to 2000 cm^{-1}.

3.4.1: Downward LWIR Emission from CO_2

The whole global warming argument is based on the effects of an increase in atmospheric CO_2 concentration on the downward LWIR flux reaching the Earth's surface. At the last glacial maximum, 20,000 years ago, the atmospheric CO_2 concentration was 200 ppm. As the Earth warmed from the Ice Age, the concentration increased to 280 ppm. This was caused by the decrease in the solubility of CO_2 in the oceans as they warmed up. As shown above in Figure 2-1, the CO_2 concentration has now reached 380 ppm. The total increase in downward LWIR flux from the most recent 100 ppm increase in CO_2 concentration is only 1.7 W.m^{-2}. The increase of 70 ppm since 1958 has contributed 1.2 W.m^{-2} to this flux. The flux increase is not distributed evenly over the CO_2 emission band, but occurs in specific spectral regions. This is shown in Figure 3-8. The spectrally resolved flux is shown for five CO_2 concentrations normalized to the 292 K black body surface emission. The increases in emission intensity in the P and R branches and in the overtones are indicated by the arrows. The change in total emission intensity is given in Figure 3-9. The calculated values are in good agreement with the 'radiative forcing constants' given by Hansen.[4] These are also plotted in Figure 3-9. It is important to note that the physical meaning of the radiative forcing constants, as an increase in downward LWIR flux is lost when they are used as empirical 'forcing constants' in the climate models.

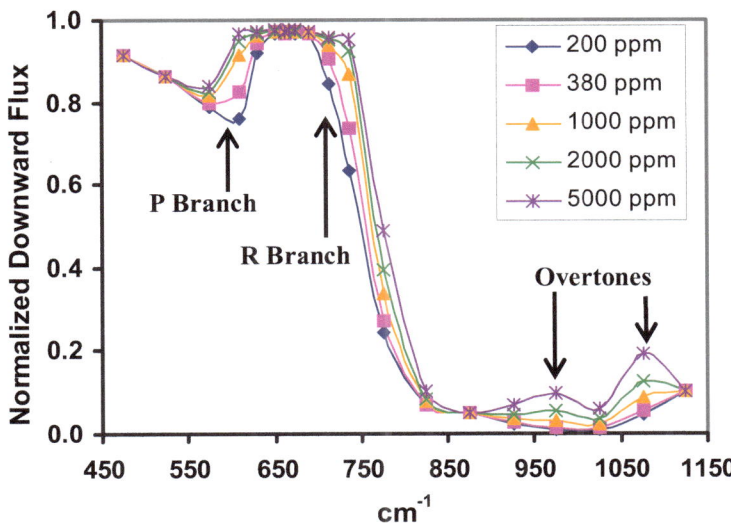

Figure 3-8: Spectrally resolved downward emission from CO_2 at the atmospheric concentrations indicated, 290 K air temperature, 292 K surface temperature, 50% RH. Normalized to the surface temperature emission.

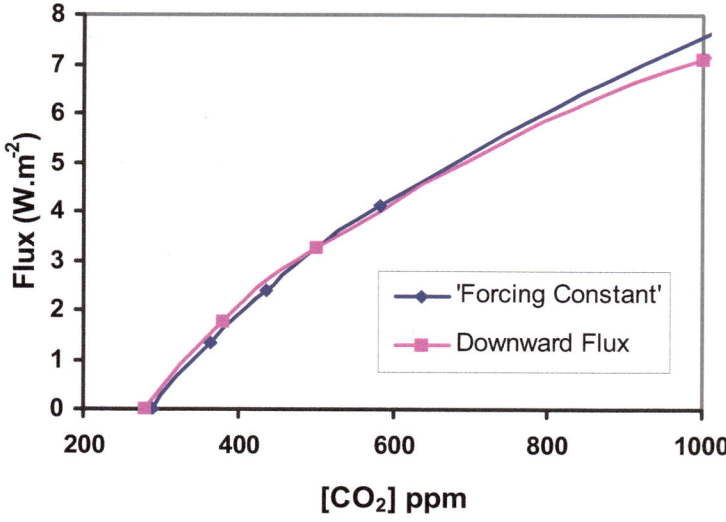

Figure 3-9: Change in total downward LWIR flux vs. CO_2 concentration from 300 to 1000 ppm.

3.5 The Lapse Rate

As air rises through the atmosphere from the surface, it expands and cools. The change in temperature with altitude is called the lapse rate. For dry air, the lapse rate is -9.8 K.km^{-1}. The air cools by almost 10 K as it ascends 1 km through the atmosphere. When the air is moist, water vapor condenses to form clouds as the air rises and cools. This releases the latent heat of evaporation of water which heats the air and reduces the lapse rate. The lapse rate is set by the surface air temperature and the humidity, although this may be modified by atmospheric circulation and mixing effects. The moist lapse rate may be as low as -3 K.km^{-1}. [28] The standard US atmosphere uses -6.5 K.km^{-1} as an average lapse rate. The air can continue to expand and cool as it rises through the troposphere until it reaches the tropopause. The US standard atmosphere places the tropopause at 11 km with a temperature of 217 K (-56 C) and a pressure of 0.22 atm. The tropopause is a rather ill defined boundary. In the tropics it can be as high as 18 km and in Polar Regions it can be as low as 8 km. The height depends on daily and seasonal surface temperatures and humidities and may change with local weather conditions at frontal boundaries. The lapse rate depends on the bulk thermodynamic properties of the atmosphere. Changes in CO_2 concentration of the order of 100 ppm can have no effect on the lapse rate or the resulting atmospheric temperature profile.

The water vapor concentration decreases by about 3 orders of magnitude upwards through the troposphere because of condensation. The CO_2 concentration only decreases by a factor of about 5 as the pressure and temperature decrease. This means that IR radiation emitted by the water vapor bands in the lower troposphere is not reabsorbed at higher altitudes. This is the main radiative cooling mechanism for the atmosphere. The calculated lapse rate and the corresponding H_2O and CO_2 concentrations for selected surface temperatures and humidities are shown in Figure 3-10 a) and b). The calculations assume adiabatic expansion for the moist air mass as it rises through the atmosphere.[29] This may be modified by local air mixing. However, Figure 3-10 clearly shows the general trends in the lapse rate and the H_2O and CO_2 concentrations. The corresponding pressure profiles are similar for all cases. The pressure decreases from 760 Torr or 1 atm. at sea level to approximately 210 Torr or 0.28 atm. at 10 km. The pressure span at 10 km is only 10 Torr for all the cases shown.

3.6 The Optical Properties of the Ocean

The energy transfer processes over land and ocean are very different. Over land, sunlight is absorbed by the surface and there is no long range subsurface thermal transport. Over the ocean, sunlight can penetrate and heat subsurface layers to depths of up to 100 m. The light is attenuated exponentially along the path length by a combination of absorption and scattering. The attenuation is wavelength dependent. Maximum optical transmission occurs in the blue-green region of the spectrum near 475 nm. Over the atmospheric spectral transmission range shown above in Figure 3-5, half of the flux is absorbed in the first meter layer of the ocean. The surface reflectivity of water is also quite low, near 2% at normal incidence. This increases as the angle of incidence increases. The change in reflectivity with angle, calculated using Fresnel's Law with a refractive index of 1.333 is shown in Figure 3-11.[30] Ocean cooling can only occur at the surface, so heat can accumulate in the subsurface ocean layers for extended periods of time and small changes in the solar flux are amplified by this accumulation process. There is no requirement that the solar flux match the cooling flux at the surface. Warm subsurface ocean layers can be recirculated and transported over long distances by wind driven ocean currents to high latitudes without significant interaction with the ocean surface. This meridional circulation is an important factor in setting the Earth's climate.

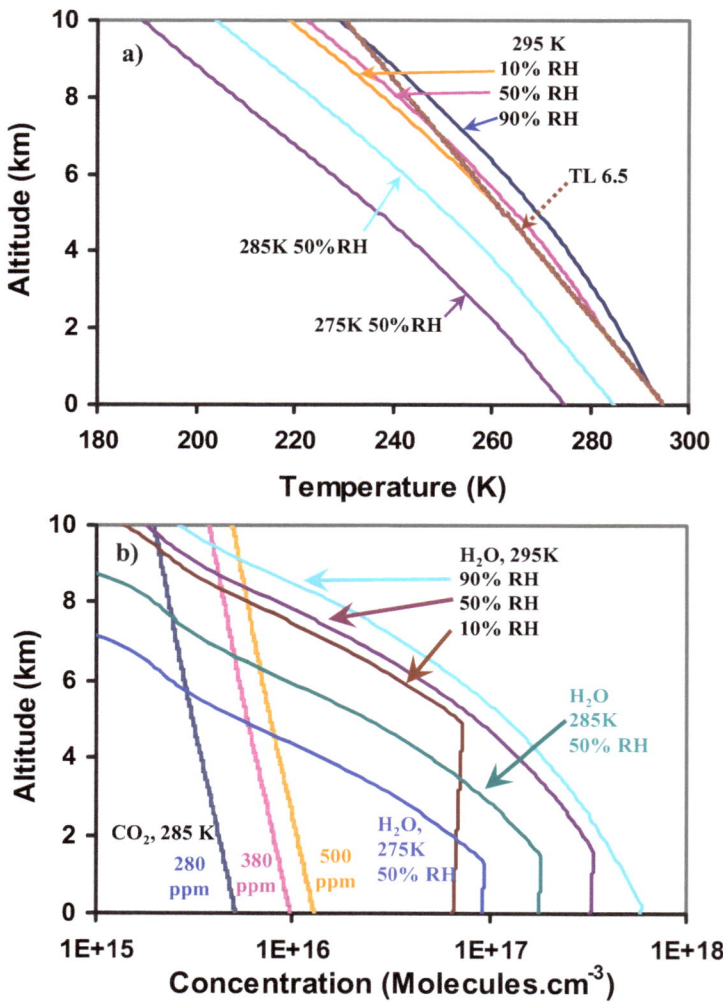

Figure 3-10: a) The lapse rate for various surface temperatures and humidities and b) the corresponding changes in H_2O and CO_2 concentration.

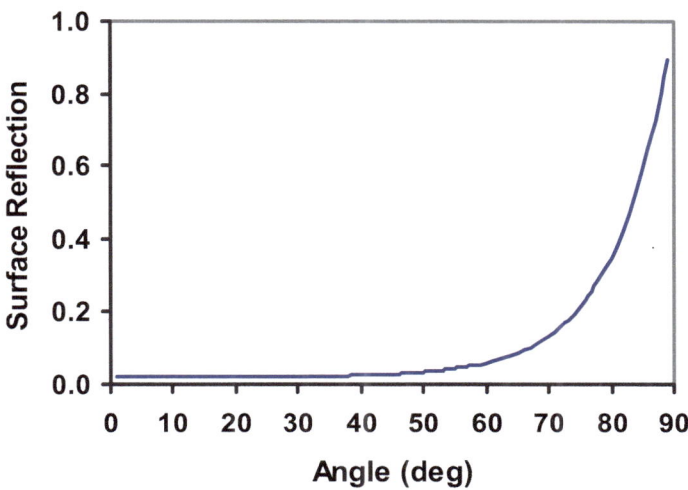

Figure 3-11: Fresnel reflection at a water surface, refractive index n = 1.333.

Although water has a high transmittance at visible wavelengths, it is almost completely opaque in the IR region. LWIR radiation cannot penetrate more than 100 μm into the ocean. This is about the width of a human hair. The surface layer of the ocean responds to an increase in LWIR radiation with an increase in temperature that converts the radiation into evaporation without any accumulation of heat into the subsurface ocean layers. These LWIR induced changes in surface temperature are hidden in the much larger wind induced changes in surface evaporation. The optical transmission (depth of 99% attenuation) of pure water is shown in Figure 3-12.[31] The spectral locations of the increases in LWIR flux from CO_2 are indicated by the arrows. The effect of wind on evaporation is considered in more detail below in Section 4.2.4.

In order to determine the total attenuation of sunlight as it is transmitted through the ocean, the spectrally resolved data from Figure 3-5 and Figure 3-12 were combined and used to calculate the integrated absorption from 300 n to 2000 nm. A wavelength independent scattering term was added to simulate real ocean waters so that the solar flux at 475 nm was attenuated to 99.9% at 100 m depth. The resulting ocean attenuation curve is given in Figure 3-13. Over half of the total flux is absorbed in the first meter layer of the ocean.

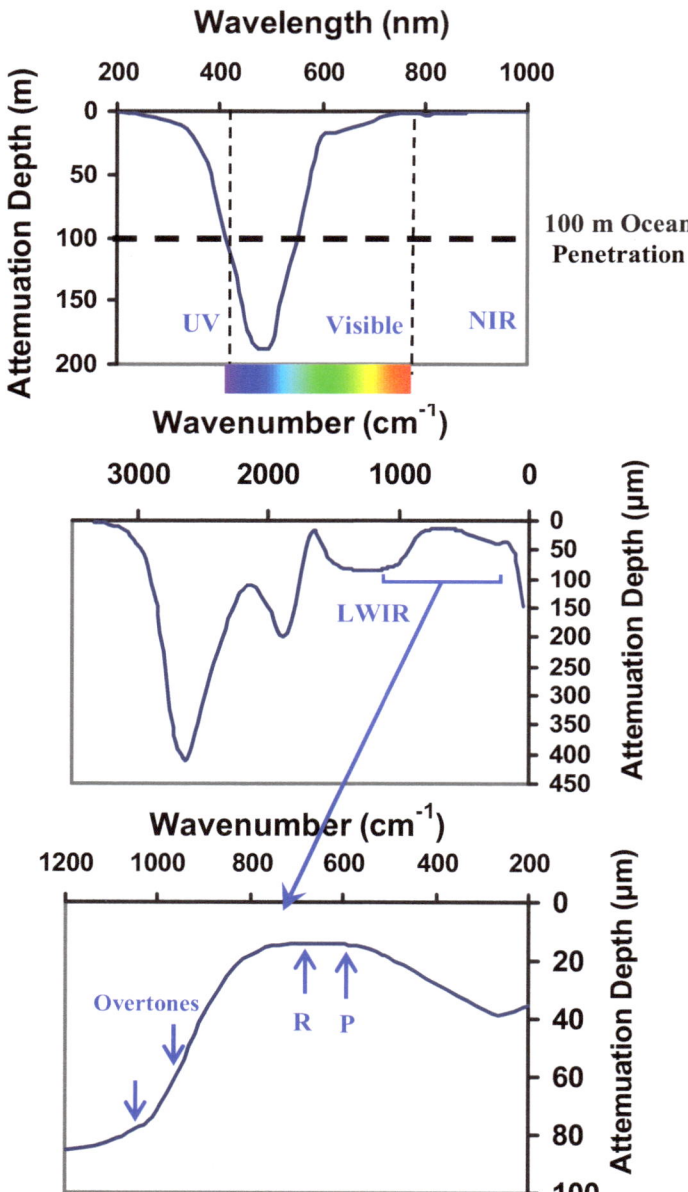

Figure 3-12: The transmission (99% attenuation depth) of pure water. Note the changes in depth units from meters to microns as the wavelength changes from the visible to the IR.

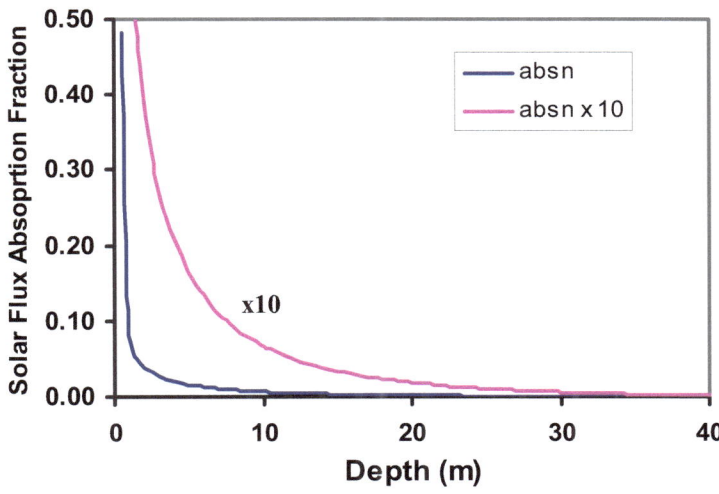

Figure 3-13: Ocean attenuation of the total solar flux, 300 to 2000 nm.

3.7 Surface Heat Transfer and Thermal Hysteresis

The Earth's surface temperature is never in equilibrium with the incident flux. Furthermore, there are relatively few measurements of the surface temperature prior to satellite observations. The concept of a long term global average surface temperature has almost no real physical meaning, although a mathematical average MSAT may be determined from the MSAT record. Such an average should be interpreted with caution and given with the error bars clearly stated. For example, Figure 3-14 shows the average daily maximum and minimum temperatures for Los Angeles Civic Center from 1921 to 2008.[32] The long term average (annual) maximum and minimum temperatures are 23.5 ±5.2 and 13.4 ±3.9 C. The errors here are the 1σ RMS deviations. Table 3-1 shows the black body flux emitted at these average temperatures, including the error bar values. The average black body flux for the maximum temperature is approximately 438 ±30 $W.m^{-2}$ and the minimum is 382 ±20 $W.m^{-2}$. These error bars are more than a factor of 10 larger than the 1.7 $W.m^{-2}$ estimated increase in LWIR flux from a 100 ppm increase in atmospheric CO_2 concentration. Furthermore, because of the 4th power dependence of the flux on the absolute temperature, the ±flux error bars are no longer equal. When the mathematical abstractions of average equilibrium temperatures are removed and replaced with real numbers, any effect from a small increase in LWIR flux is lost in the noise of the errors of the averaging process. Figure 3-14 also shows that the surface temperature peak is reached in middle of August, almost 2 months after the summer solstice peak in solar flux. This clearly shows the temperature hysteresis associated with the ground thermal storage. Similar thermal hysteresis effects are also found in the daily temperature changes.

Table 3-1

	T C	T K	Flux (W.m^{-2})	1σ Flux
Max+1σ	28.7	301.7	469.9	31.6
Max	23.5	296.5	438.4	
Max-1σ	18.3	291.3	408.4	-30.0
Min+1σ	17.3	290.3	402.8	21.2
Min	13.4	286.4	381.6	
Min-1σ	9.5	282.5	361.3	-20.4

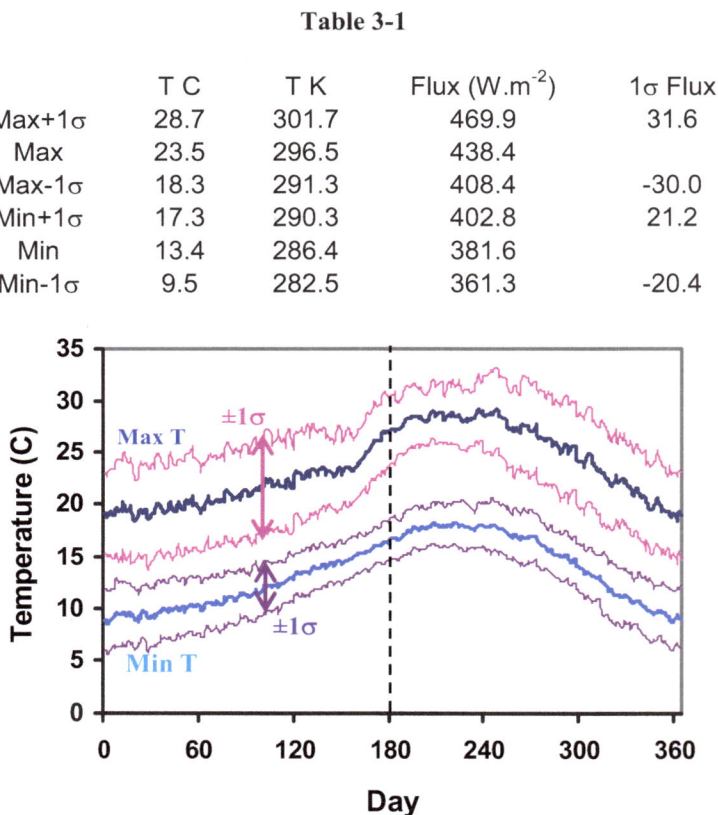

Figure 3-14: Maximum and minimum daily temperatures for the Los Angeles Civic Center, 1921 to 2008.

Figure 3-15 shows the daily temperature variation recorded by the author in the Los Angeles area from June 21 to 30, 2008. Figure 3-16 shows the daily temperature for June 22 on an enlarged scale. These figures show the approximate surface temperature for dry soil and the air temperature 0.75 and 1.5 m above the ground. The surface temperature was recorded using a thermistor placed just below the soil surface. During the day, the surface temperature is significantly warmer than the air temperature. The heat storage induced thermal hysteresis is also clearly visible. The dip in the surface temperature in the middle of the day is caused by shading of the ground thermometer by the legs of the air thermometer enclosure support structure. There was also some shading from surrounding structures at the beginning and end of the day. The assumption that there is a black body equilibrium temperature at the surface has no basis in physical reality. The surface temperature can only be determined from the short term flux balance using a realistic heat transfer model that includes subsurface thermal conduction.

It is also important to note that the night time ground and air temperatures are coupled. The ground continues to cool by convection until the air and ground temperatures are similar. When the convection is reduced, the dominant cooling term becomes the emission of LWIR radiation from the surface through the LWIR window. The air may now cool by transferring heat to the ground. This depends on the local wind speed and air turbulence. When the air temperature is reduced below the dew point, condensation occurs and latent heat effects also have to be included. As the air temperature changes with the local weather systems, the ground temperature also changes.

Figure 3-15 and Figure 3-16 illustrate these general trends. For example, both the minimum ground and air temperatures decreased by 5 C during the first 3 days shown in Figure 3-15. The daytime increase in surface temperature caused by solar heating of the surface can clearly be seen. This drives the daily convection. The slower night time cooling is also clearly visible. Thermal hysteresis effects are also observed in ocean heating. Neither the daily nor the seasonal temperature peaks correspond to the peak in the solar flux. Such hysteresis effects are usually neglected in discussions of the Earth's surface temperature, but they are a fundamental part of the surface energy transfer. They provide a more stringent validation test of any thermal model than simple temperature prediction.

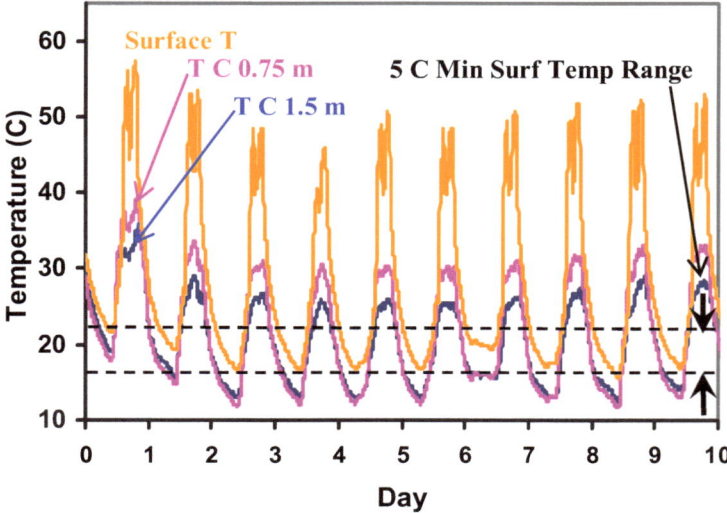

Figure 3-15: Daily surface and 0.75 and 1.5 m air temperatures, June 21 to
30 2008, dry soil, full sun, Los Angeles area.

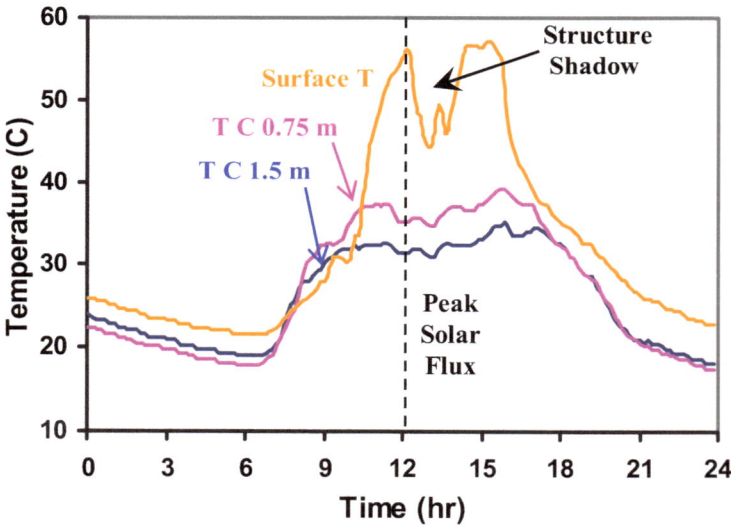

Figure 3-16: Daily surface and air temperatures, June 21 2008, dry soil,
full sun, Los Angeles area.

3.8 The Solar Heating of an Earth without an Atmosphere

Various 'what if' arguments have been proposed to illustrate aspects the greenhouse effect. These include a water free or greenhouse gas free atmosphere and an atmosphere free 'bare rock' Earth. The 'bare rock' case provides a good illustration of the errors made in the simple equilibrium assumptions. The surface temperature may be calculated quantitatively using a simple finite element thermal conduction model with time dependent solar illumination.[3,33] The 'bare rock' Earth rotates with the correct axial tilt of 23.5° and the solar flux varies with latitude and season. The basic thermal model is illustrated in Figure 3-17.

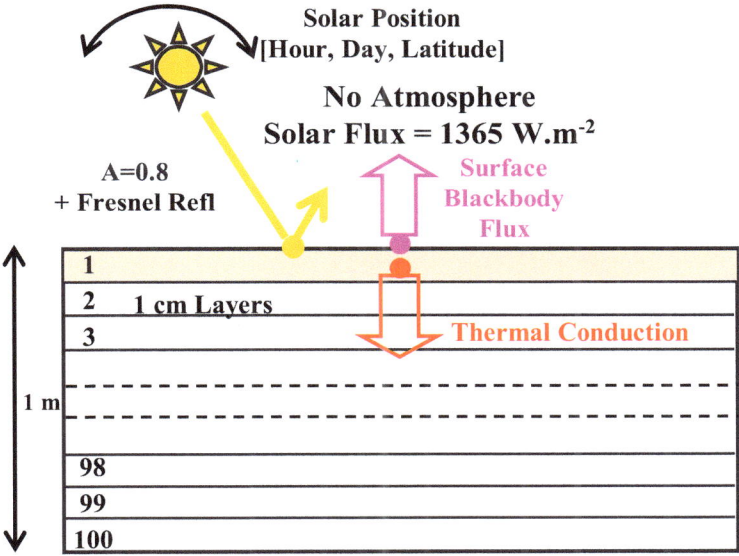

Figure 3-17: The 'Bare Rock' surface heating model.

The solar flux is set to 1365 W.m^{-2}. There is no atmosphere. The surface thermal properties are those of dry sand. The sun heats the surface during the day and part of the heat is conducted below the surface. The surface cools continuously by blackbody emission. The surface absorption for the solar flux is set to 0.8. An angular dependent Fresnel reflection for a refractive index of 1.5 is also included at the surface. The surface acts as a pure blackbody radiator for the LWIR emission. The finite element model uses 1 cm layers in a finite element central difference thermal conduction algorithm. The time step is set to 1 minute and the output is usually sampled at 2 hour intervals. For single day, time dependent analysis, the full 1 min. time resolution is used. The daily maximum and minimum temperatures are also extracted from the surface temperatures. The solar flux is calculated for the time of year, time of day and latitude with the Earth tilted at 23.5° to the solar plane. This flux is coupled to the first 1 cm layer at the surface. The black body thermal flux and the flux change due to thermal conduction in or out of the surface are calculated. The net flux change in the first 1 cm layer is used to calculate the new surface temperature.

Using this model, the surface temperatures at 2 hour intervals for a full year were calculated at latitudes of 0, 15, 30, 45 and 60°. In addition, the full daily surface temperature variation at 1 minute intervals was output for days 80, 171, 265 and 356 corresponding to the equinox and solstice points. The calculated maximum and minimum daily temperatures as a function of latitude are shown in Figure 3-18 and Figure 3-19. A second plot with an enlarged temperature scale is also shown in each Figure. The first point that needs to be made is there is no such thing as an average temperature, except as a mathematical abstraction. The maximum daily temperatures vary from 95 to -162 C and the minimum temperatures vary from -64 to –179 C.

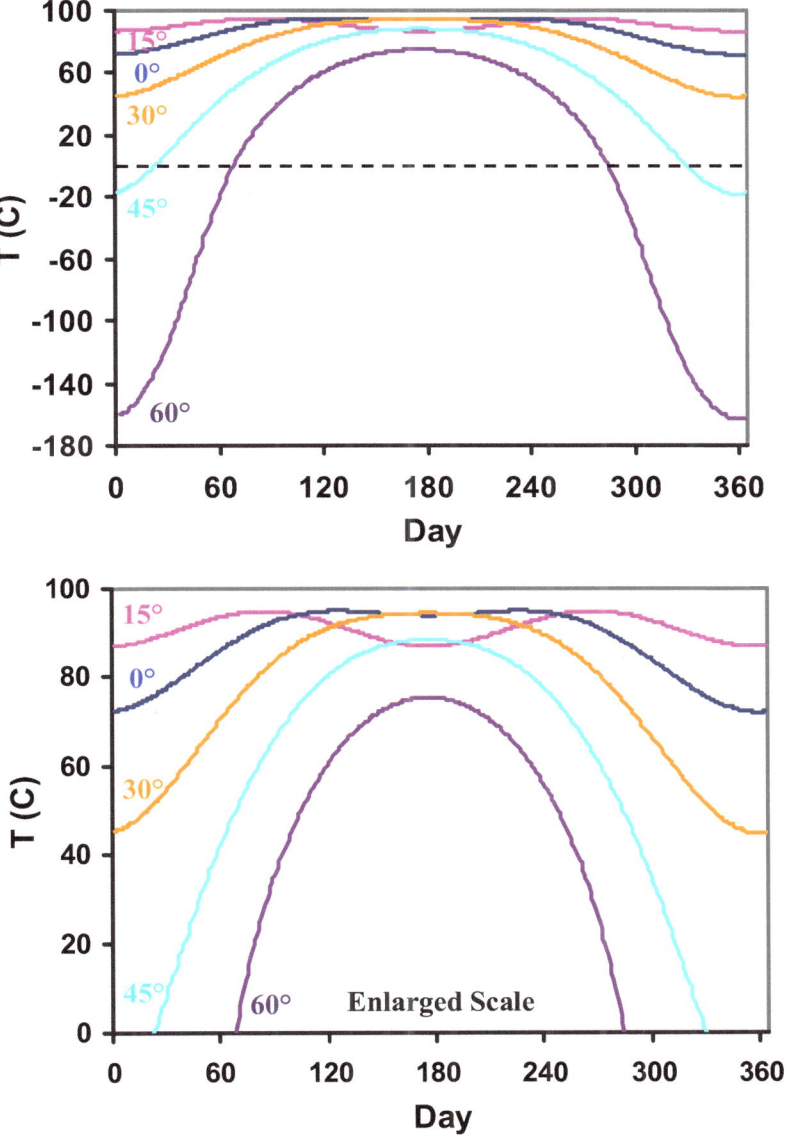

Figure 3-18: Maximum daily temperatures at latitudes shown.

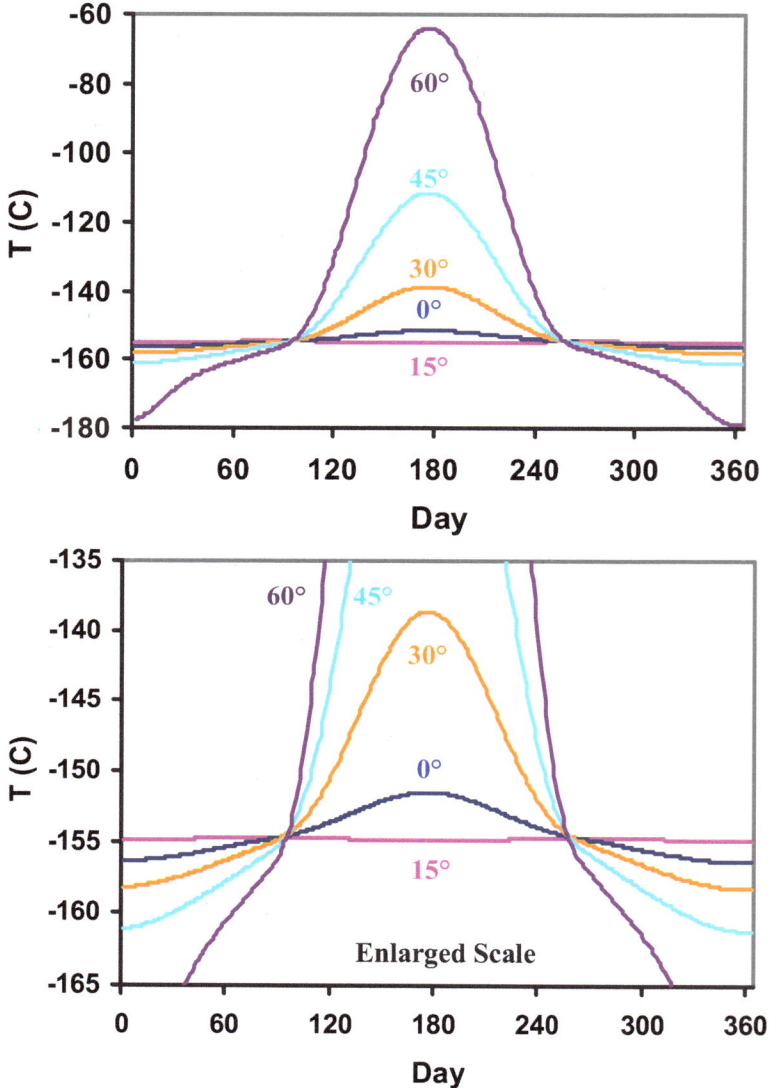

Figure 3-19: Minimum daily temperatures at latitudes shown.

The daily surface temperature changes for the four days, 80, 171, 265 and 356 that correspond to the solstice and equinox points are shown in Figure 3-20 through Figure 3-24 for the 5 latitudes used in the calculations. These plots are not symmetric about the solar flux, but show the characteristic time lag of a thermal storage reservoir. It is also important to emphasize that the surface is not in thermal equilibrium with the solar flux. The temperature derived from the solar flux is not the same as the surface temperature. In general the surface temperature will be lower because of the heat transfer to the subsurface thermal reservoir. In addition, there is a time delay or phase shift between the peak solar flux and the peak surface temperature. This is illustrated in Figure 3-25. The upper plot shows the solar flux and surface temperature for day 80 (vernal equinox) at a latitude of 30°. The center plot shows the surface temperature compared to the temperature derived from the incident solar flux. The temperature derived from the average solar flux is also shown (flat line). The lower plot shows the phase shift between the solar flux temperature and the surface temperature. The phase difference is approximately 20 minutes. The solar flux temperature has the same phase as the solar flux. It is also important to note that different averaging techniques give significantly different average surface temperatures. The average temperature can be derived from the average absorbed solar flux, the time series average of the surface temperature or the averages of just the maximum and minimum temperatures. The average of the actual absorbed flux, the average flux temperature, the time series average surface temperature, the minimum and maximum surface temperatures and the (Min+Max)/2 temperature average as a function of latitude are given in Figure 3-26 through Figure 3-29 for the four equinox/solstice days.

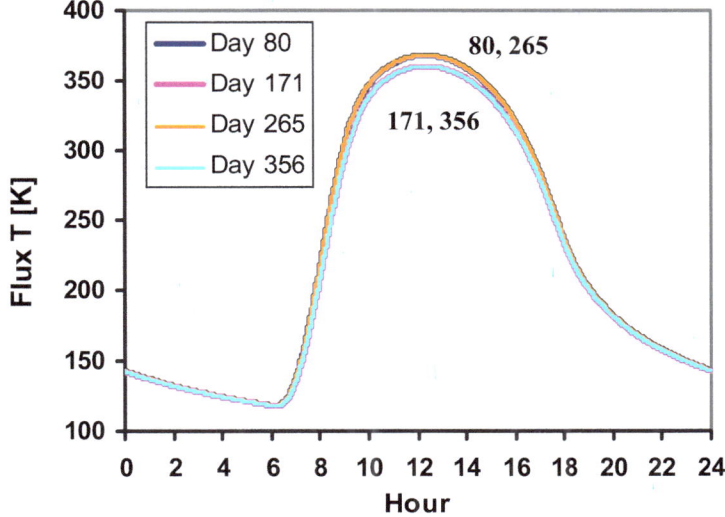

Figure 3-20: Daily surface temperature variation for the solstice and equinox points at 0° latitude (equator). The plots for days 80 and 265 (equinox) and for days 171 and 356 (solstice) almost overlap.

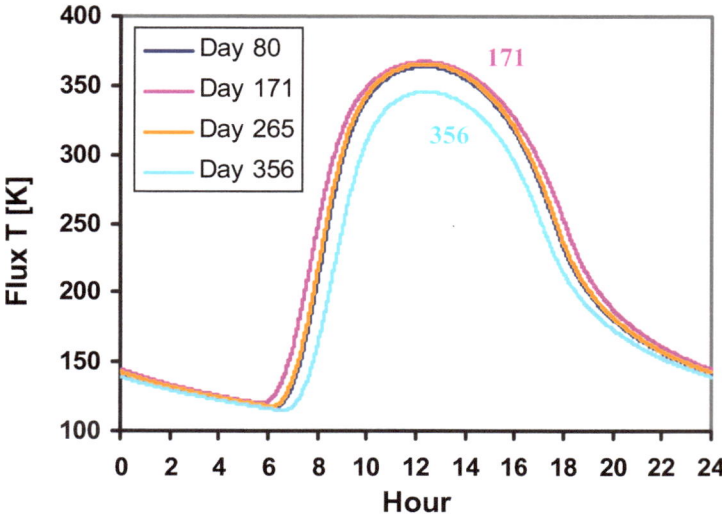

Figure 3-21: Daily surface temperature variation for the solstice and equinox points at 15° latitude. The plots for days 80 and 265 almost overlap.

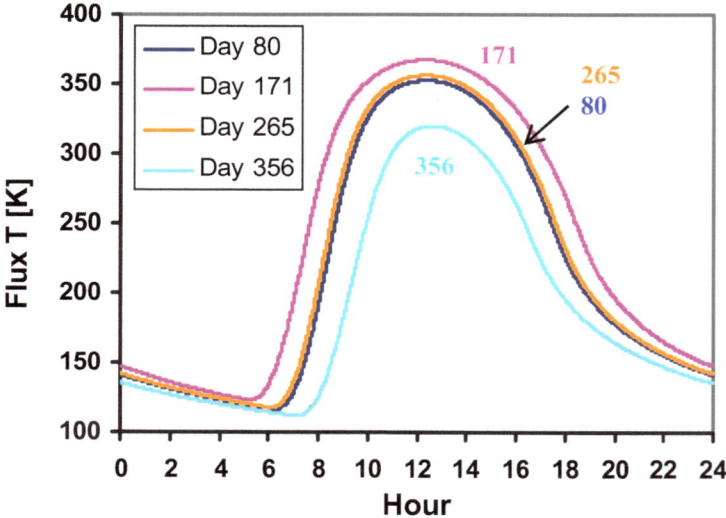

Figure 3-22: Daily surface temperature variation for the solstice and equinox points at 30° latitude. The plots for days 80 and 265 almost overlap.

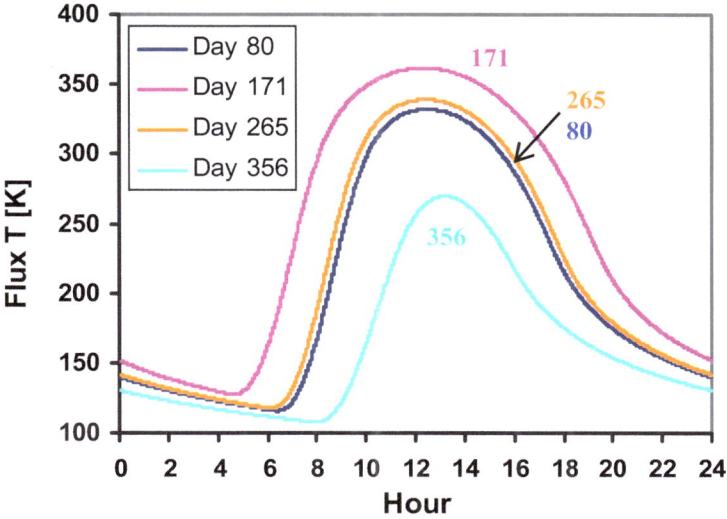

Figure 3-23: Daily surface temperature variation for the solstice and equinox points at 45° latitude.

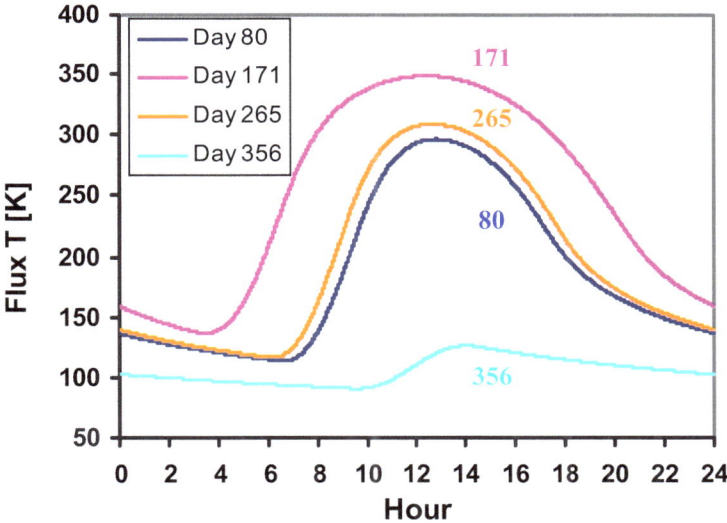

Figure 3-24: Daily surface temperature variation for the solstice and equinox points at 60° latitude.

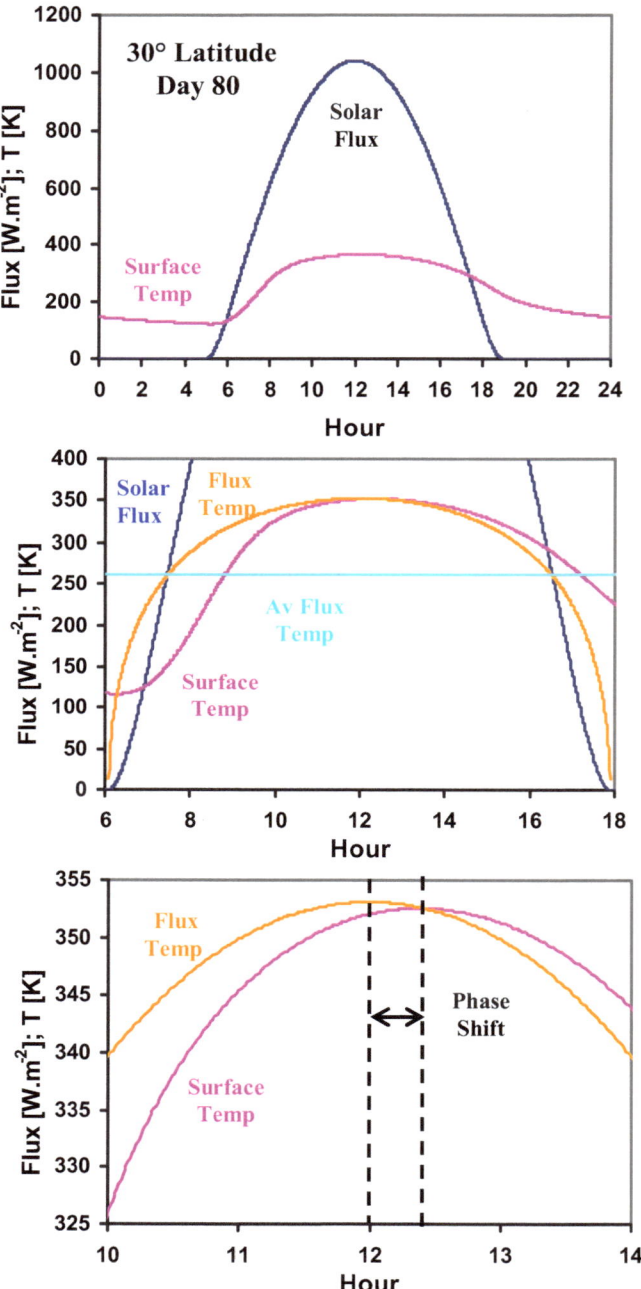

Figure 3-25: Surface temperature detail for day 80, vernal equinox, at 30° latitude. For discussion see text.

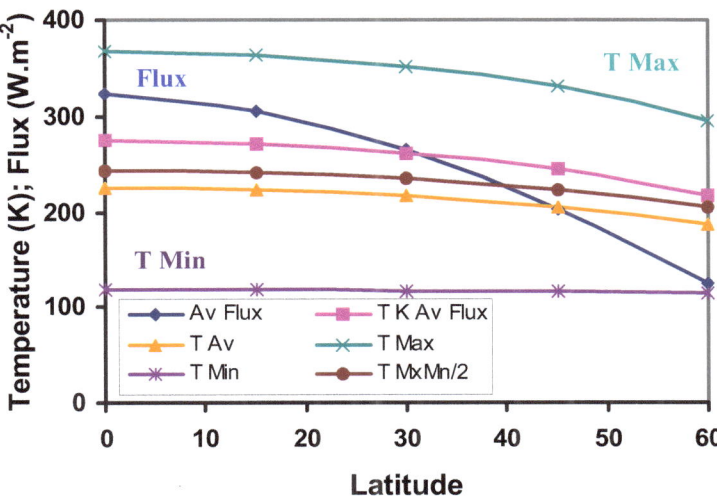

Figure 3-26: Flux, flux av., min, max, and mxmn/2 derived temperature averages vs. latitude for day 80.

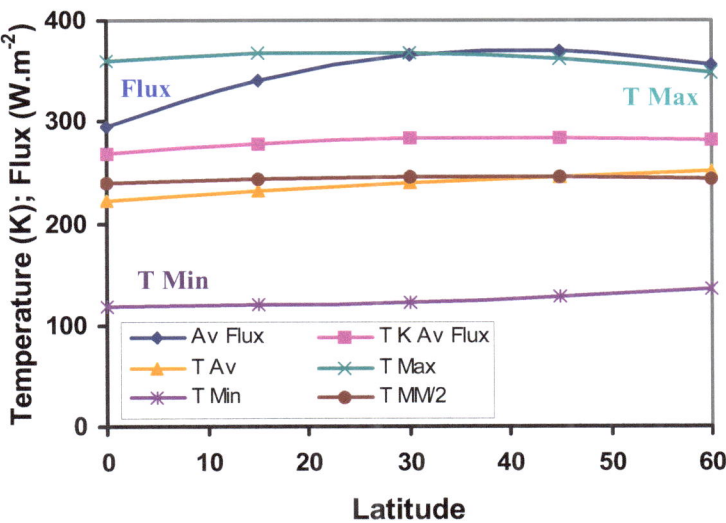

Figure 3-27: Flux, flux av., min, max, and mxmn/2 derived temperature averages vs. latitude for day 171.

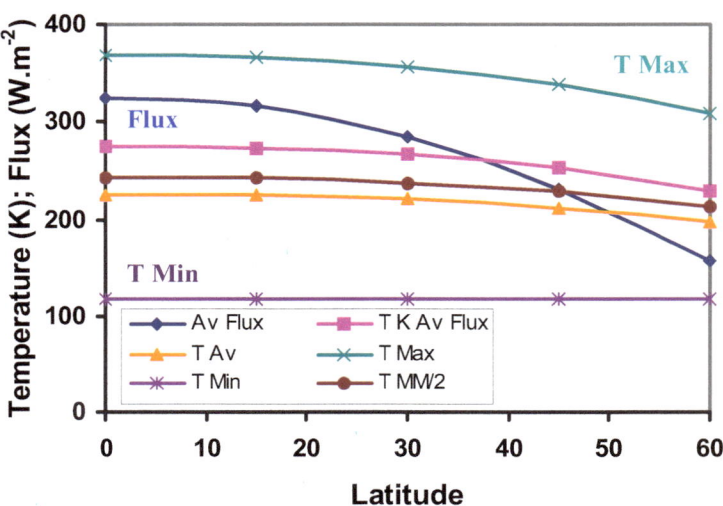

Figure 3-28: Flux, flux av., min, max, and mxmn/2 derived temperature averages vs. latitude for day 265.

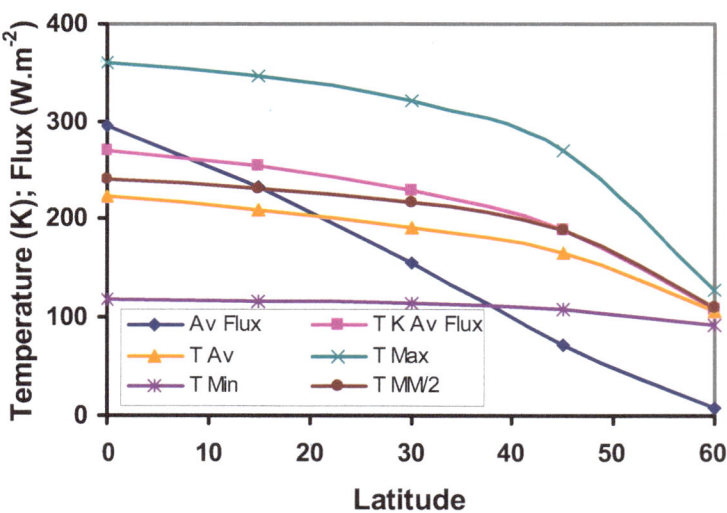

Figure 3-29: Flux, flux av., min, max, and mxmn/2 derived temperature averages vs. latitude for day 356.

A defining characteristic of a thermal storage reservoir with a varying heat load is the time delay or phase shift between the input flux and temperature response. The phase shifts in minutes of the peak surface temperature relative to the input flux (noon) are shown in Figure 3-30. The lower plot shows the data on an expanded time scale. At low latitudes and high solar flux, the phase shift is approximately 20 minutes. This increases at low flux and high latitudes until it reaches 120 minutes for the 60° latitude winter solstice case. There is also a difference in the peak temperature derived from incident the solar flux and the actual peak surface temperature. This is shown in Figure 3-31. The lower plot shows the temperature differences on an enlarged scale. The peak surface temperature is always lower than the solar flux derived temperature because of the thermal storage. This means that the concept of an 'average equilibrium surface temperature' used in radiative forcing calculations has no basis in physical reality. The calculations used to derive the results presented in this section had a 1 minute time resolution. This was the time step interval needed for accuracy in the finite element analysis. The 'bare rock' Earth model also illustrates the climate averaging paradox. The long term averages have to be calculated from the short term data. There is no simplified 'equilibrium average' climate that can be determined by other means.

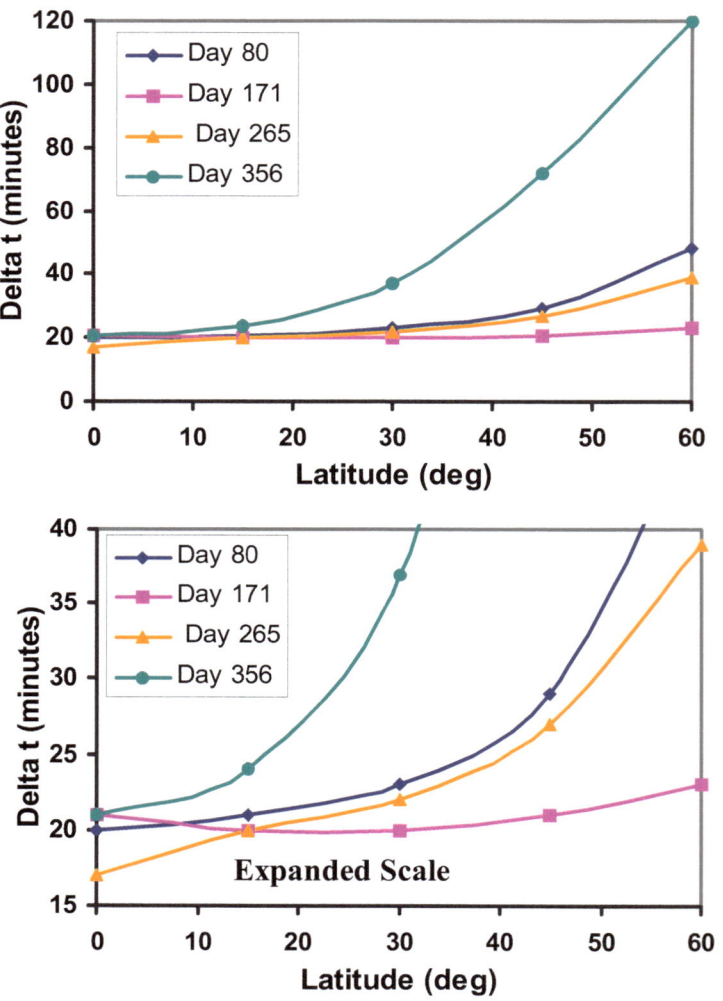

Figure 3-30: Phase shift between the peak solar flux and the peak surface temperature.

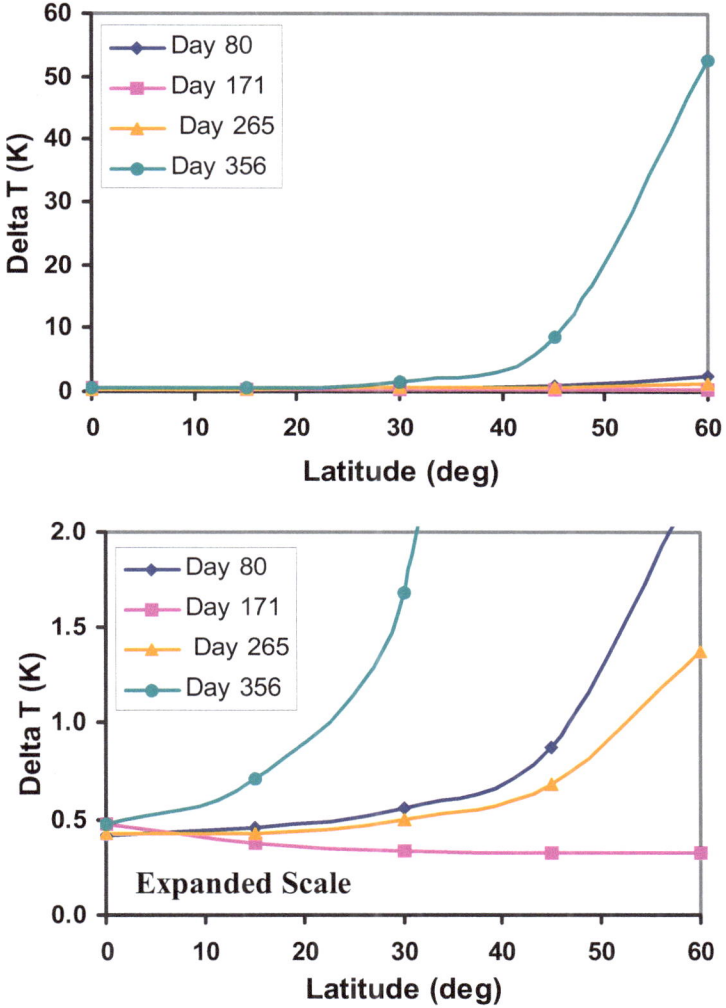

Figure 3-31: Difference in peak temperature between the solar flux temperature and the surface temperature.

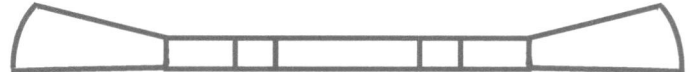

4.0 DYNAMIC ENERGY TRANSFER

As explained above in Chapter 1, there are six interacting energy transfer processes that combine dynamically to determine the Earth's climate. The total flux is dynamically coupled to four different thermal reservoirs, the oceans, the land and the lower and upper troposphere. The starting point is to consider the energy balance at the surface that determines the surface temperature. This controls the subsequent energy transfer through the atmosphere. Since the air-ocean and air-land interfaces behave differently, particularly with regard to subsurface heating and transport, these have to be considered separately.

For a time interval Δt, a formal surface energy balance may be written as:

$$\Delta Q_{sun} = \Delta Q_{sens} + \Delta Q_{lat} + \Delta Q_{ir} + \Delta Q_{ssh} \qquad (4.1)$$

where ΔQ_{sun} is the solar heating flux, ΔQ_{sens} is the sensible heat, or dry air convection, ΔQ_{lat} is the latent heat flux or water evaporation, ΔQ_{ir} is the net LWIR emission from the surface and ΔQ_{ssh} is the subsurface thermal transport. In a conventional radiative forcing analysis, the time interval is set to 1 second and a long term average energy balance is assumed between the solar flux and the net LWIR emission. The convection, latent heat and subsurface heat transfer are ignored. This approach has no basis in physical reality. The surface temperature is determined by the short term balance between all of the terms in Equation 4.1.

The incident solar flux depends upon the position of the sun in the sky and the aerosol content of the atmosphere along the solar optical path to the surface. Under full summer sun conditions, this flux may reach or exceed 1000 W.m^{-2}. This corresponds to a black body emission temperature of 91.4 C. Clouds and fog can block the sun completely. Some of the solar flux is reflected at the surface. This depends upon the nature of the surface and the angle of incidence. For smooth dielectric surfaces, the reflectivity may be calculated from the Fresnel equations using the (complex) refractive index. The change in reflectivity for water, $n = 1.333$, with no absorption, is shown above in Figure 3-11. The solar flux and the refractive index are wavelength dependent. The spectral distribution of the solar spectrum is shown above in Figure 3-5.

The sensible heat or dry air convection depends upon the temperature difference between the surface and the air above. It is also dependent on the wind speed at the surface. As the cooler air from above mixes with the warm air heated by the surface, convective eddy currents with a wide range of length and time scales are formed. The length scale increases with height above the surface. The measurement and analysis of the details of convective mixing is complex and simple bulk transport coefficients are generally used. The surface convective flux is commonly measured using eddy covariance techniques in which the air velocity components are determined using acoustic methods. Independent measurements of water vapor and carbon dioxide concentrations using IR absorption are used to derive the latent heat flux and vegetation respiration from the eddy covariance data.[34]

Over the oceans, wind driven evaporation is usually the dominant cooling process. It increases rapidly with ocean surface temperature. On a molecular scale, evaporation requires the breaking of the hydrogen bonds between liquid water molecules and the transport of the separated vapor molecules away from the surface by convection. The latent heat of water is large, about 2.5×10^3 MJ.m^{-3}. For the oceans, the evaporation rate depends upon the water vapor concentration (humidity) gradient near the surface, the wind speed and the surface turbulence. Over land, it also depends on the diffusion of moisture from the soil or vegetation. The land evaporation rate also increases rapidly with temperature.

The net IR emission from the surface depends upon the surface and air temperatures, the humidity and the aerosol content or cloud cover. It is determined by the balance between the upward LWIR flux from the surface and the downward LWIR flux from the atmosphere. Under clear sky conditions, when the surface and air temperatures are similar, the net surface LWIR cooling flux is approximately 40 W.m^{-2} or 0.14 MJ.m^{-2}.hr^{-1}. Under low humidity conditions, this may increase to 100 W.m^{-2} and under low cloud or fog conditions, the flux may be zero.

The surface heating effects are very different at the ocean and land interfaces. In the ocean, the solar flux is attenuated exponentially along the optical path, depending on the local absorption and scattering coefficients. Surface cooling produces a surface or skin layer about 1 mm thick that is cooler than the bulk water layer underneath. This cooler water sinks and is replaced by warmer water from below. At night, a diurnal mixing layer with a uniform temperature is established. The depth of this layer depends on the thermal gradients and the magnitude of the surface cooling. The ocean and air temperatures near the surface are usually similar. The average surface air temperature is typically 2 C cooler than the corresponding ocean temperature. Over land, the solar flux is absorbed at the surface and heat is conducted below the surface by the subsurface thermal gradient that is established. The surface temperature can exceed 50 C for a dry surface under full summer sun illumination. Most of the heat that is stored during the day is released in the evening, but there is a long term seasonal warming and cooling as the solar flux changes.

It is also important to emphasize that the surface temperature needed for the dynamic energy balance is the actual surface temperature. The meteorological surface air temperature (MSAT) is the air temperature measured in an enclosure placed at eye level, 1.5 to 2 m above the ground. At night, the surface temperature and MSAT are often similar. However, during the day, the land surface temperature may easily be 20 C warmer than the MSAT.

4.1 The Air Land Interface

Over land, the solar flux is absorbed at the surface and there is no long range thermal transport. As discussed in the previous section, the difference between the surface temperature and the MSAT must be properly accounted for in the surface temperature analysis. The surface temperature is the temperature of the ground under our bare feet. The meteorological surface air temperature (MSAT) is the air temperature measured in an enclosure located for convenience at eye level, 1.5 to 2 m above the ground. There is no simple or obvious relationship between the real surface temperature and the MSAT.[3,35,36] In order to investigate the energy transfer processes that determine the surface temperature and the MSAT, an AmeriFlux monitoring site, operated by the University of California, Irvine, located in Limestone Canyon Regional Park, east of Irvine was selected for analysis.[37] Data for the year 2008 was chosen for this study. The complete data set consisted of half hour averages of 17 parameters: friction velocity; air temperature; wind direction; wind speed; CO_2 flux; H_2O flux; sensible heat flux; latent heat flux; CO_2 concentration; H_2O concentration; incoming photosynthetic active radiation;, reflected photosynthetic active radiation; incoming global solar radiation; reflected global solar radiation; relative humidity; precipitation and net radiation.

Figure 4-1 shows the daily maximum and minimum measured air temperatures for 2008. The 8 day maximum and minimum surface (skin) temperatures from satellite data are also shown. The minimum air and surface temperatures are similar, but the maximum surface temperature during summer is approximately 15 C higher than the measured maximum air temperature. This clearly shows that MSATs are not the same as actual ground surface temperatures. At this particular monitoring site, there were well defined fluctuations in air temperature and humidity that were related to the shift from the ocean to the desert origin of the local weather system. Under ocean influences, temperatures were lower, the humidity was higher, and night and early morning clouds could develop. Under desert influences, the temperatures were higher and the humidity was lower. This site also experienced well known Santa Ana wind conditions when air that originates from the inland high desert regions is adiabatically compressed and produces very hot and dry conditions with very strong local winds. Such adiabatic compression can increase air temperatures by 10 C.

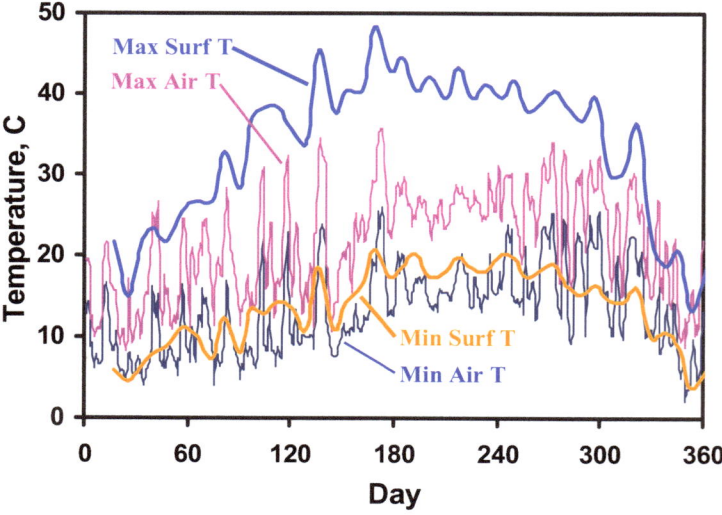

Figure 4-1: Maximum and minimum air and surface temperatures for 2008 measured at the UC Irvine 'Grasslands' Ameriflux site.[37] **Air temperatures are daily min and max values. Surface temperatures are 8 day average satellite data. Average maximum surface temperatures are ~12 C warmer than the average maximum air temperatures. The air temperature 'spikes' are from offshore desert winds, heated by adiabatic compression (Santa Ana winds).**

In order to examine the effect of solar heating and latent heat flux on the daily temperature rise, the total daily surface heating flux (absorbed solar flux minus the daylight latent heat flux, $MJ.m^{-2}.day^{-1}$) and the daily air temperature rise were plotted and scaled to overlap. This is shown in Figure 4-2. The total surface heating flux (thick line) shows the characteristic seasonal variation with a summer peak. Superimposed on this are decreases due to periods of cloud cover. The latent heat flux is derived from eddy covariance measurements that are not recorded during periods of rainfall. The daily temperature excursions generally follow the trend of the surface heating flux but with large fluctuations because of the transitions between ocean and desert influences. Figure 4-3 shows the daily total latent heat flux ($MJ.m^{-2}.day^{-1}$) for the daytime and night-time periods. Most of the flux is associated with the solar heating during the day. The seasonal peak in the latent heat flux occurs in the spring following the winter rainfall. It should also be noted that in the near IR region, the solar flux can heat and evaporate water by direct absorption in addition to the thermal heating of the surface. Because of the large fluctuations in the daily temperature excursions between ocean and desert conditions at this site, it was not possible to examine the effects of changes in temperature caused by humidity induced variations in LWIR absorption.

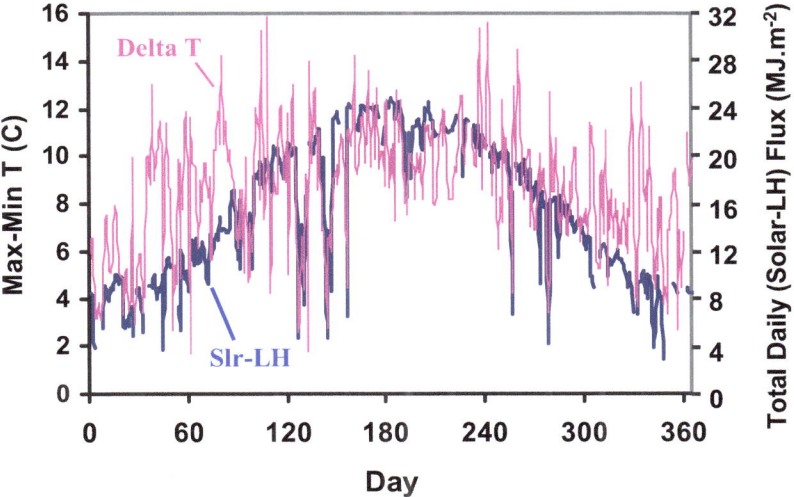

Figure 4-2: Total daily (Solar-LH) flux (MJ.m^{-2}.day^{-1}) and daily air temperature increase (Max-Min) for 2008.

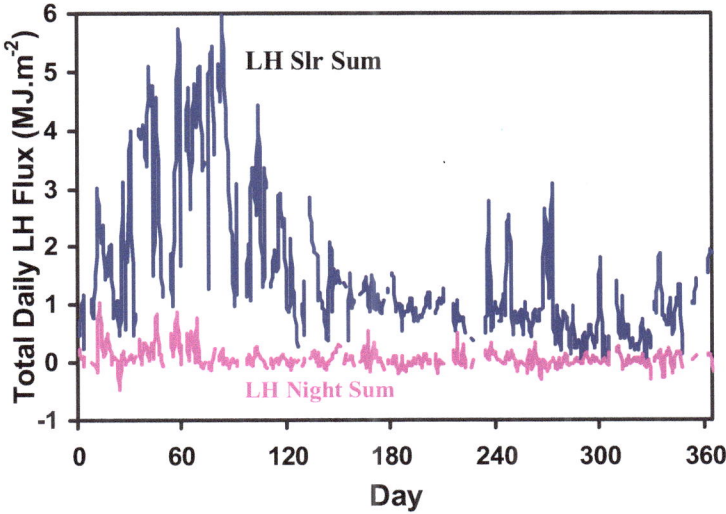

Figure 4-3: Daily latent heat flux totals (MJ.m^{-2}) for daytime (sunlight) and night time evaporation.

Surface heat transfer is illustrated in Figure 4-4. This is based on the measured flux terms from the Ameriflux site for June 30, 2008 and the calculated surface temperature and subsurface heating terms obtained using the model described below in Section 4.2.1. During the day, the ground is heated by the sun. The daily difference between the minimum and maximum surface air temperature depends on the absorbed solar flux at the surface, the sensible heat flux, or convective cooling, the latent heat flux, the water vapor concentration and any change in bulk air temperature of the air mass of the weather system. Under full summer sun conditions, the short term dry surface temperature can easily exceed 50 C. However, the increase in blackbody LWIR flux as the surface temperature increases from 20 to 50 C is only 200 $W.m^{-2}$. This means that most of the solar flux is coupled back into the atmosphere by convection, not thermal radiation. When the surface is moist, some of the surface heat is removed by evaporation. This latent heat flux reduces the surface temperature and the latent heat is released into the atmosphere through cloud formation, usually at altitudes above 1 km. This latent heat release adds to the convection at higher altitudes.

The surface heating also establishes a subsurface thermal gradient that conducts heat below the surface. This stored heat is released as the ground cools during the late afternoon and evening and adds to the surface convection. At night, the air and the ground cool to similar temperatures, convection slows considerably and the surface cools mainly through LWIR emission to space via the atmospheric LWIR transmission window. The typical LWIR flux is ~40 $W.m^{-2}$. Under low humidity conditions, this may increase to 100 $W.m^{-2}$. Downward LWIR emission from low cloud cover may close the LWIR window and balance all of the upward LIWR flux from the surface. The diurnal change in magnitude of the flux terms is so large that an increase of 1.7 $W.m^{-2}$ in downward LWIR flux from a 100 pm increase in atmospheric CO_2 concentration can have no effect on the surface temperature. This is considered in more detail in the next Section.

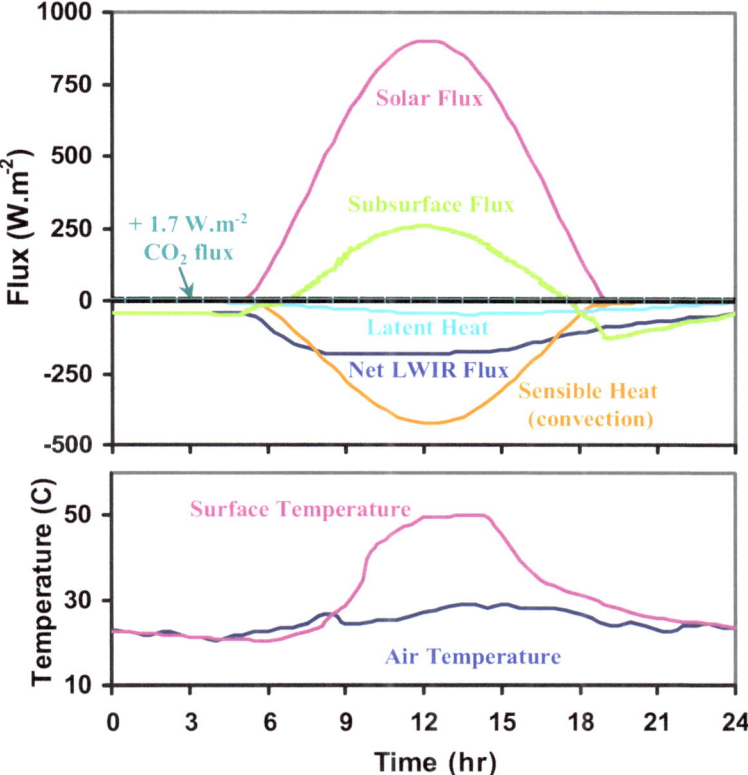

Figure 4-4: Surface energy transfer: a) Surface flux terms and b) surface and air temperatures. Data based on UC Irvine 'Grasslands' AmeriFlux site data recorded June 30 th 2008.[37] Subsurface flux and surface temperature are calculated from a surface heating model developed using the 2008 site data. The addition of +1.7 W.m⁻² LWIR flux from a 100 ppm increase in atmospheric CO_2 concentration can have no measurable effect on the surface temperature. See reference 35 for further discussion.

4.1.1 Simulation of the Land Heating

The land heating model described above in Section 3.8 was modified to simulate smoothed surface temperatures derived from Figure 4-1. The downward atmospheric LWIR emission and the latent and sensible heat cooling fluxes were coupled to the first 1 cm layer. The absorbed solar flux was calculated using the IEEE 738 solar flux algorithm with an absorption fraction of 0.8 and an angle dependent external Fresnel reflection for a refractive index of 1.5.[30] The absorption fraction and the Fresnel term were derived from an analysis of the measured solar flux data.

The latent heat flux was calculated as a fraction of the absorbed solar flux. This fraction was determined from the total daytime latent heat shown in Figure 4-3 and the total daily absorbed solar flux from Figure 4-2. It was incorporated into the model as a third order polynomial in the day number using the Trendline Algorithm in Excel™. The IR emission of the surface was calculated from Stefan's law using the surface temperature calculated by the heat transfer model. The minimum air temperature was calculated using a trendline polynomial fit to the minimum surface temperature from Figure 4-1. The minimum air and surface temperatures are similar and a fit to the minimum surface temperature had a higher correlation coefficient than the air temperature (0.86 compared to 0.42). A daytime temperature cycle was superimposed on this using a fixed value of 8 C scaled by the cosine of the solar zenith angle. The 8 C term was the annual average of the daily temperature rise. The IR emission of the air was calculated using Stefan's law applied to the air cycle temperature. A constant term of 43 $W.m^{-2}$ was used for the LWIR window loss, derived from the annual average of the net night time IR flux. A constant convection loss coefficient of 21 $J.m^{-2}.C^{-1}.s^{-1}$ was used to calculate the convection loss from the surface-air temperature difference. The heat conducted into the surface was calculated by the 2D thermal conduction model. The published thermal properties of dry sand were used in the initial calculations. The energy balance of these seven flux terms was used to calculate the new surface temperature of the top 1 cm layer of the surface conduction model. This surface temperature was then used as input to the thermal conduction model. A time step of 1 minute was used. The model is illustrated in Figure 4-5.

As the surface heats up during the day from the solar flux, some of this heat is accumulated by the subsurface layers and released during the night. There is also a seasonal variation in the subsurface heat content. No equilibrium assumption is made. The latent heat flux is derived from the measured heat flux at the monitoring site. Using this approach with a simple physical model, reasonable agreement between the calculated surface temperatures and the fit to the measured satellite surface temperatures is obtained. This is illustrated in Figure 4-6.

The Dynamic Greenhouse Effect

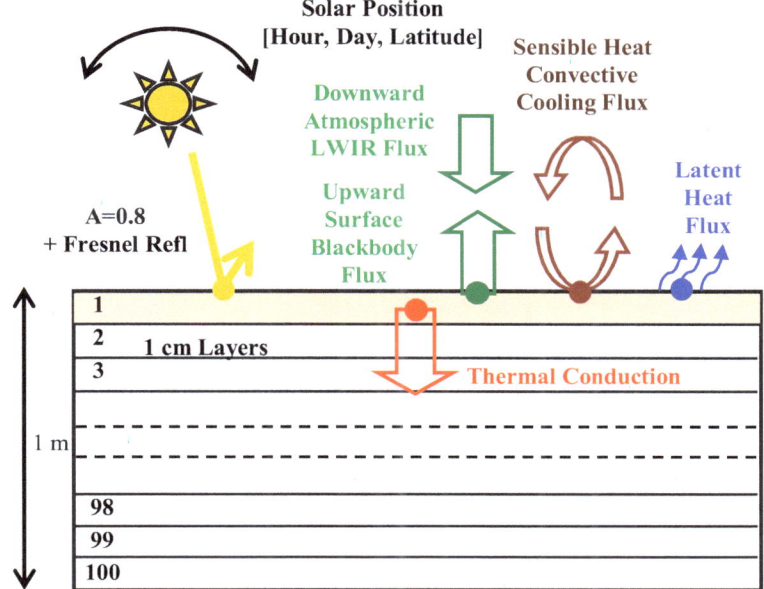

Figure 4-5: Thermal model used to simulate the land surface temperatures from Figure 4-1.

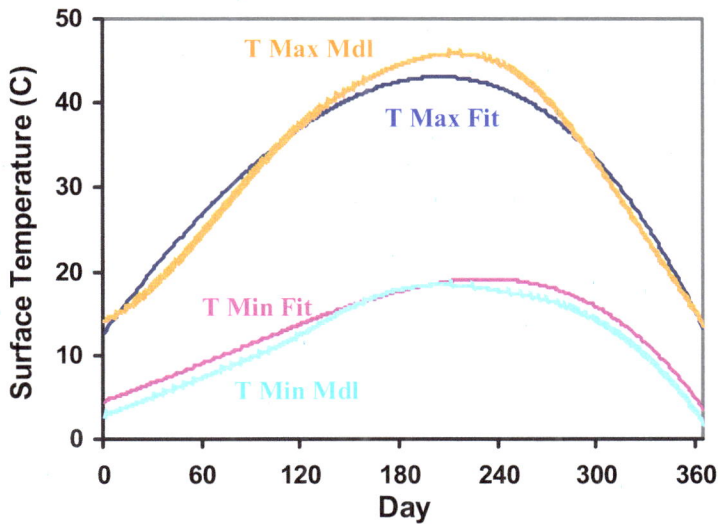

Figure 4-6: Calculated maximum and minimum temperatures compared to smoothed satellite surface temperature measurements from Figure 4-1.

49

The effect of an increase of 1.7 W.m^{-2} in the downward LWIR flux on the simulated surface temperatures was investigated by decreasing the IR cooling term in the model from 43 to 41.3 W.m^{-2}. This corresponds to an increase in atmospheric CO_2 concentration of 100 ppm. The average annual increases in the calculated minimum and maximum surface temperatures were 0.065 and 0.062C. The daily differences are shown in Figure 4-7. All of the daily changes were between 0.060 and 0.067 C. The reason that these temperature changes are so small is because the change in LWIR flux is a small fraction of the daily heat flux dynamically coupled to the surface. A flux of 1.7 W.m^{-2} corresponds to a daily cumulative flux of 0.15 MJ.m^{-2} that is superimposed on a total daily flux in excess of 10 MJ.m^{-2}. Because of the large daily fluctuations in the actual maximum and minimum temperatures and in the solar, IR and latent heat flux terms, the small changes shown in Figure 4-7 are not detectable in the surface temperature record. This result clearly demonstrates that any equilibrium assumption used to calculate the surface temperature is invalid. The change in heat flux must be coupled dynamically using a heat conduction model with realistic heat flux terms and surface thermal properties. Once the surface temperature is calculated from the dynamic flux balance, it is clear that there can be no CO_2 induced global warming.

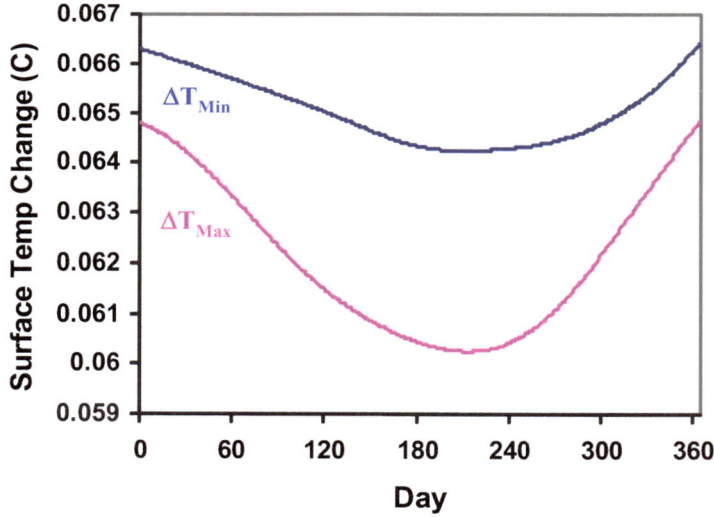

Figure 4-7: Effect of a 1.7 W.m^{-2} increase in downward LWIR flux on the calculated daily maximum and minimum surface temperatures.

4.1.2 The Ocean Influence on the Minimum MSAT

Based on the results presented above, there can be no CO_2 'global warming signature' in the MSAT record. Instead, the observed changes in the MSAT record can be explained as a combination of changes in ocean surface temperatures, urban heat island effects and [fraudulent] 'adjustment' or 'homogenization' of the climate record. This may be understood by examining the minimum MSAT record in regions that are influenced by ocean temperatures. The minimum MSAT is an approximate indicator of the bulk surface air temperature of the local weather system. The daily increase in temperature from minimum to maximum is an indicator of the solar surface heat load that is coupled by convection to the surface air layer. Since 75% of the Earth is ocean, most weather systems are formed over the ocean. Any long term change in the MSAT record may be expected to contain information on the change ocean surface temperatures along the path of the prevailing weather systems. This effect may be studied by comparing the minimum MSAT record of a selected weather station to the appropriate ocean surface temperature record over the same time period. The magnitude of the daily increase in MSAT is related to the solar flux as shown above in Figure 4-2. For the State of California, and neighboring regions, the appropriate reference is the Pacific Decadal Oscillation, PDO and for the UK and surrounding regions, the appropriate reference is the local Atlantic Multidecadal Oscillation, AMO. Figure 4-8 shows the 5 year rolling average of the PDO from 1904. The data were downloaded from the University of Washington website.[38] The long term trend line for the full data set is almost flat. This was obtained using the linear trendline algorithm in Excel™. However, over shorter time periods, a significant slope, positive or negative may be derived from the data. The California weather station data was downloaded from the Western Region Climate Center website and used 'as received'.[39] The data for Pierce College was downloaded from the college website.[40]

Figure 4-9 shows the minimum MSAT for Los Angeles Civic Center, five year rolling average from 1925 to 2005. The PDO from Figure 4-8 is also shown over the same period of record. The linear trend lines for both data sets from 1925 to 2005 are also shown. The minimum MSAT data for the LA Civic Center shows the characteristic 'signature' of the PDO superimposed on an upward sloping baseline. The difference in slope between the PDO and the weather station data, 0.022 $C.yr^{-1}$ is an approximate indicator of the urban heat island effect for Los Angeles.

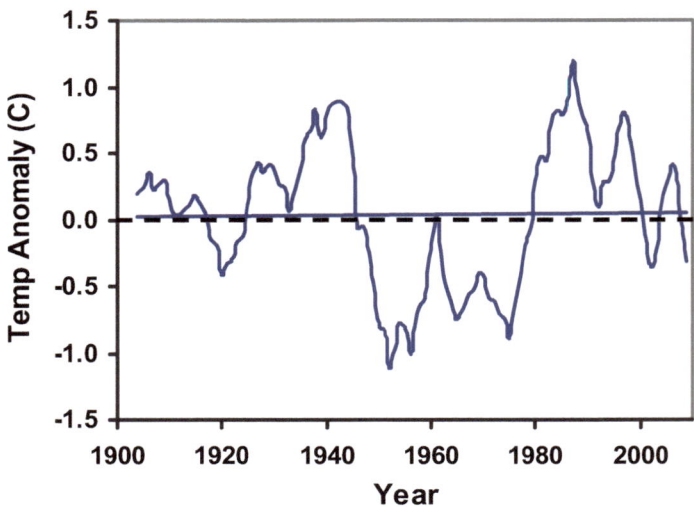

Figure 4-8: Pacific Decadal Oscillation (PDO), five year rolling average from 1904.

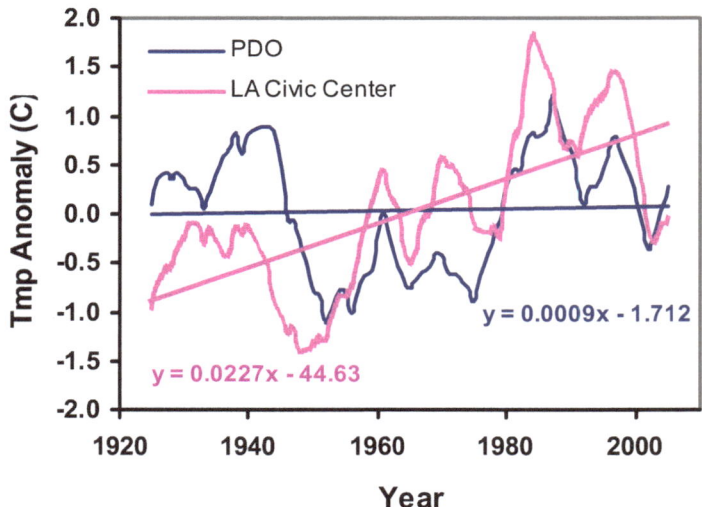

Figure 4-9: Minimum MSAT temperature, 5 year rolling average, for the LA Civic Center from 1925 to 2005. The PDO and the trend lines over the same time period are also shown.

Figure 4-10 shows the minimum MSAT for Los Angeles Airport, LAX, from 1950 to 2008 with the PDO and trend lines over the same time period. In this case, the slope of the station data is close to that of the PDO. The slope difference is 0.005 C.yr^{-1}. LAX is located on the coast, approximately 25 km west of the Civic Center. The marine layer and onshore flow at LAX significantly reduce the urban heat island effect compared to the Los Angeles Civic Center. Figure 4-11 shows the minimum MSAT for Nevada City, from 1935 to 2009, again with the PDO reference and trend lines. This station is located about 100 km NE of Sacramento. It should be a 'rural' station with a small urban heat island effect. Inspection of the data reveals a temperature rise of over 3 C between 1978 and 1990. This was probably caused by a change in station location. The temperature trends on either side of this anomaly are 0.06 and 0.05 C.yr^{-1}. These are still high for a 'rural' station, indicating that further investigation is needed.

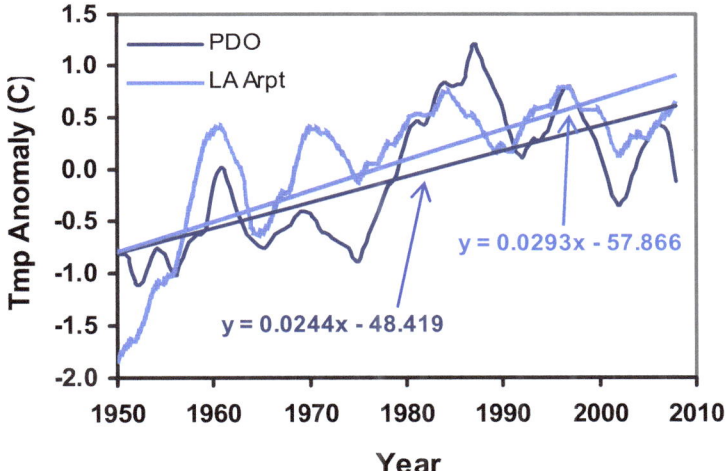

Figure 4-10: Minimum MSAT temperature, 5 year rolling average, for LA Airport from 1950 to 2008. The PDO and the trend lines over the same time period are also shown.

The minimum MSAT trend data for LA Civic Center, LAX and Nevada City are examples of a general technique that compares the minimum MSAT data to a reference set of ocean surface temperatures along the approach path of the prevailing weather systems. While care is needed in the interpretation of such data, the difference in slope between the ocean reference and the station data is a measure of the local urban heat island effect on the station. In addition, obvious discrepancies such as steps or unexpected peaks in the station data can be used to flag data anomalies for further investigation.

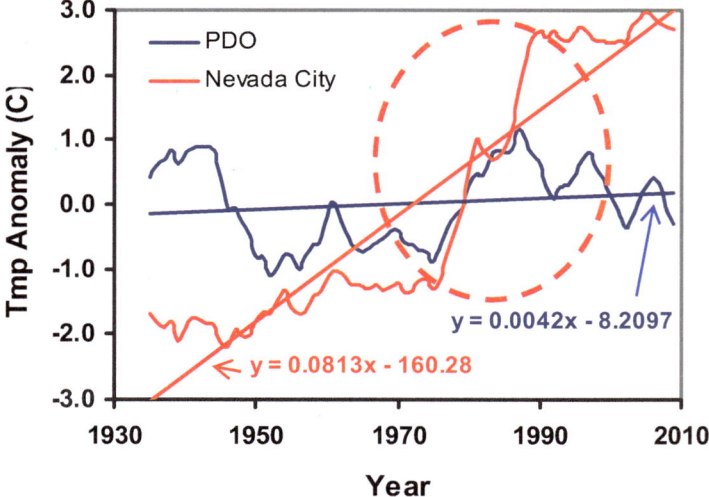

Figure 4-11: Minimum MSAT temperature, 5 year rolling average, for Nevada City from 1935 to 2009. The PDO and the trend lines over the same time period are also shown. The circled temperature anomaly was probably caused by a station relocation.

Using this technique, a total of 34 California weather stations were analyzed. Further details are given in Reference 36. Stations with a minimum record duration of 50 years were selected to be representative of the full geographical and climate extent of California. The linear trend data are plotted in Figure 4-12. The stations were divided into four groups. The first group was 'coastal' which included 10 coastal weather stations from Crescent City to San Diego. The second group was 'rural' which included 9 stations with warming trends below 0.01 C.yr^{-1}. These were mainly located in rural areas. The third group was 'urban' which included 14 stations with warming trends above 0.01 C.yr^{-1}. The fourth group was 'anomalous' where visual inspection of the station data indicated obvious discrepancies associated for example with changes in location, that require further investigation. In most cases, the anomaly only impacted part of the data set and the rest of the data could be processed normally with a reduced time scale. Numbers after the station name indicate truncated, reprocessed data sets.

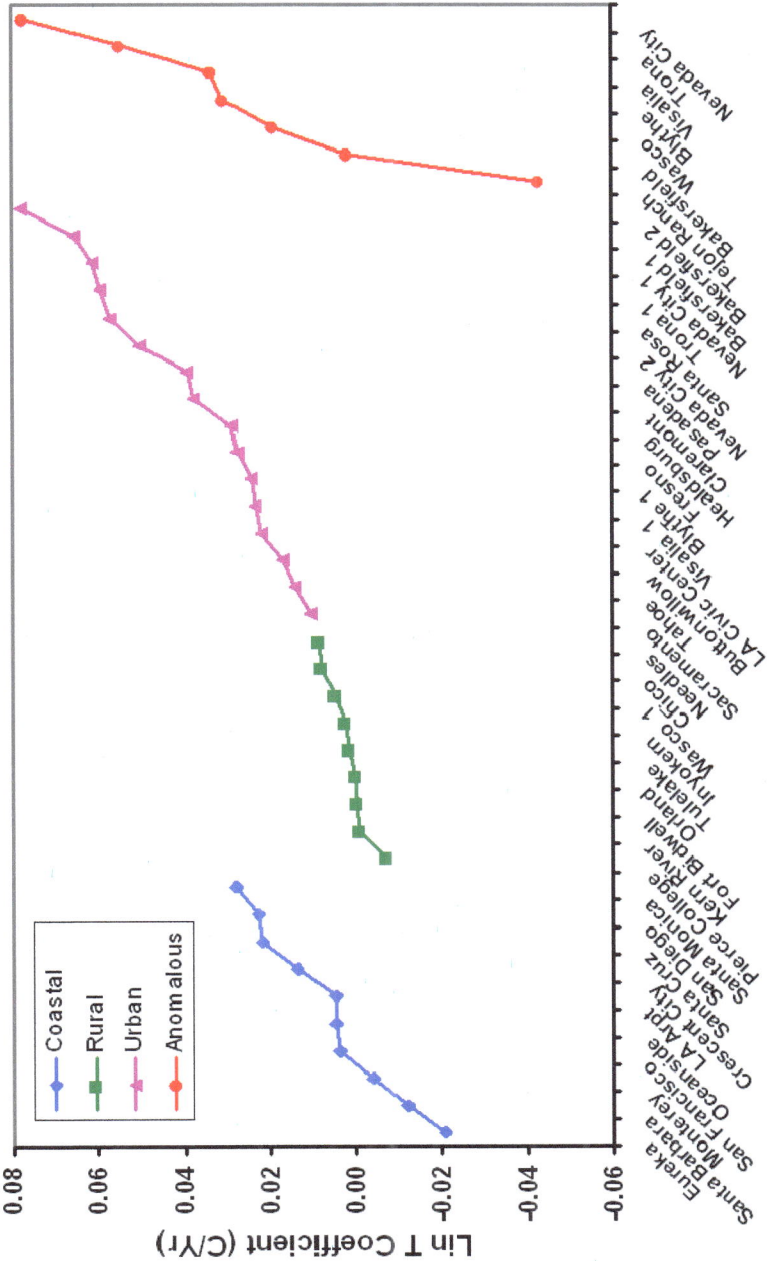

Figure 4-12: Linear warming trend data for the California weather stations. The stations were divided into four groups based on location and linear trend magnitude.

These results show that the climate of the State of California, as measured by the minimum MSAT weather station record is set mainly by the PDO. Superimposed on the PDO is an approximately linear urban heat island effect that depends on the local microclimate of the individual station and the influence of urban development on diurnal and seasonal subsurface heat storage. There is no evidence of CO_2 induced global warming in the California data. Each station has its own unique microclimate and local bias effects. The 'one size fits all' practice of averaging station data into 5° latitude and longitude 'boxes' overestimates climate change by adding urban heat island effects to the natural climate trends.

The analysis of the minimum MSAT record for the 34 California weather stations was extended to 33 UK weather stations using the local Atlantic Multi-decadal Oscillation as the ocean surface temperature reference. The weather station and AMO data were downloaded from the Hadley Center web site and used 'as received'.[41,42] The rolling 5 year average of the 45-50 N, 10-15 W; 45-50 N, 5-10 W and 50-55 N, 10-15 W 5 x 5° 'boxes' from the HadSST2 database was used as the local AMO reference. Figure 4-13 shows the five year rolling average of the AMO used in the UK station analysis. Figure 4-14 shows the linear trend analysis for Heathrow, which had the largest urban heat island effect in the UK station data. Figure 4-15 shows the linear trend data for the 33 stations. These were divided into three groups based on the magnitude of the linear trend. Group 1 contained four stations with slightly negative trends (-0.005 to 0.0 C.yr^{-1}). Group 2 contained 21 stations with trends between 0 and 0.01 C.yr^{-1} equivalent to the 'rural' California stations. Group 3 contained the remaining 8 stations with trends above 0.01 C.yr^{-1}, equivalent to the California 'urban' stations. There is no evidence of CO_2 induced global warming in the UK data.

Overall, the linear trends for the UK stations showed lower urban heat island effects compared to the California stations. The UK linear trend range was from approximately -0.005 to +0.025 C.yr^{-1}. The California linear trend range was from approximately -0.02 to +0.08 C.yr^{-1} (excluding the anomalous station data). The urban heat island effect is a measure of the increase in heat stored in the ground (and structures) as a result of urban development. It depends on both the solar heating of the area and the changes in latent heat flux as a result of urban run off and vegetation loss. Since the UK receives less sunshine and more rainfall than most of California, the urban heat island effects are lower for the UK. The general climate trend for the UK is decreasing rainfall and increasing sunshine from N to S and W to E. This is reflected in the urban heat island trends.

The higher trend values tend to be located in the SE. However, each station has its own microclimate that needs to be evaluated on a case by case basis. The concept of 5 x 5° 'boxes' of climate averages is not a very useful one and leads to overestimates of climate warming from local urban heat island effects.

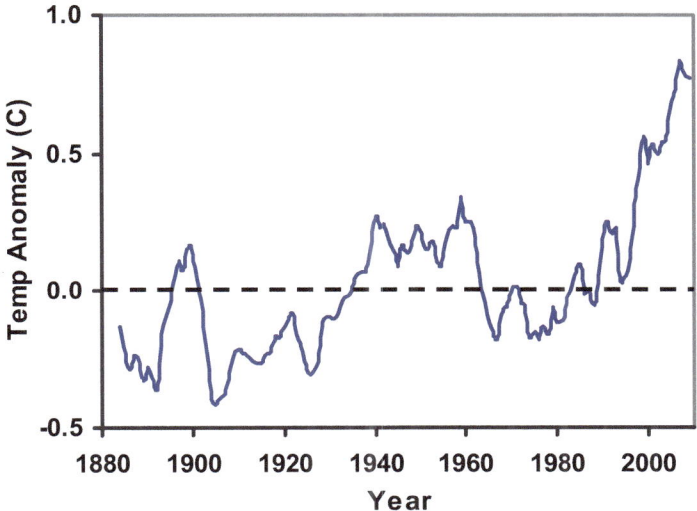

Figure 4-13: Local AMO reference, 5 year rolling average used in the analysis of the UK weather stations.

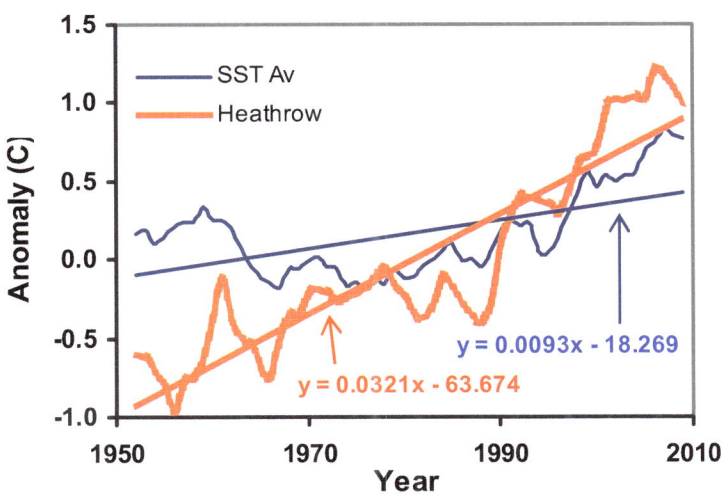

Figure 4-14: Linear trend analysis for Heathrow using the AMO as reference.

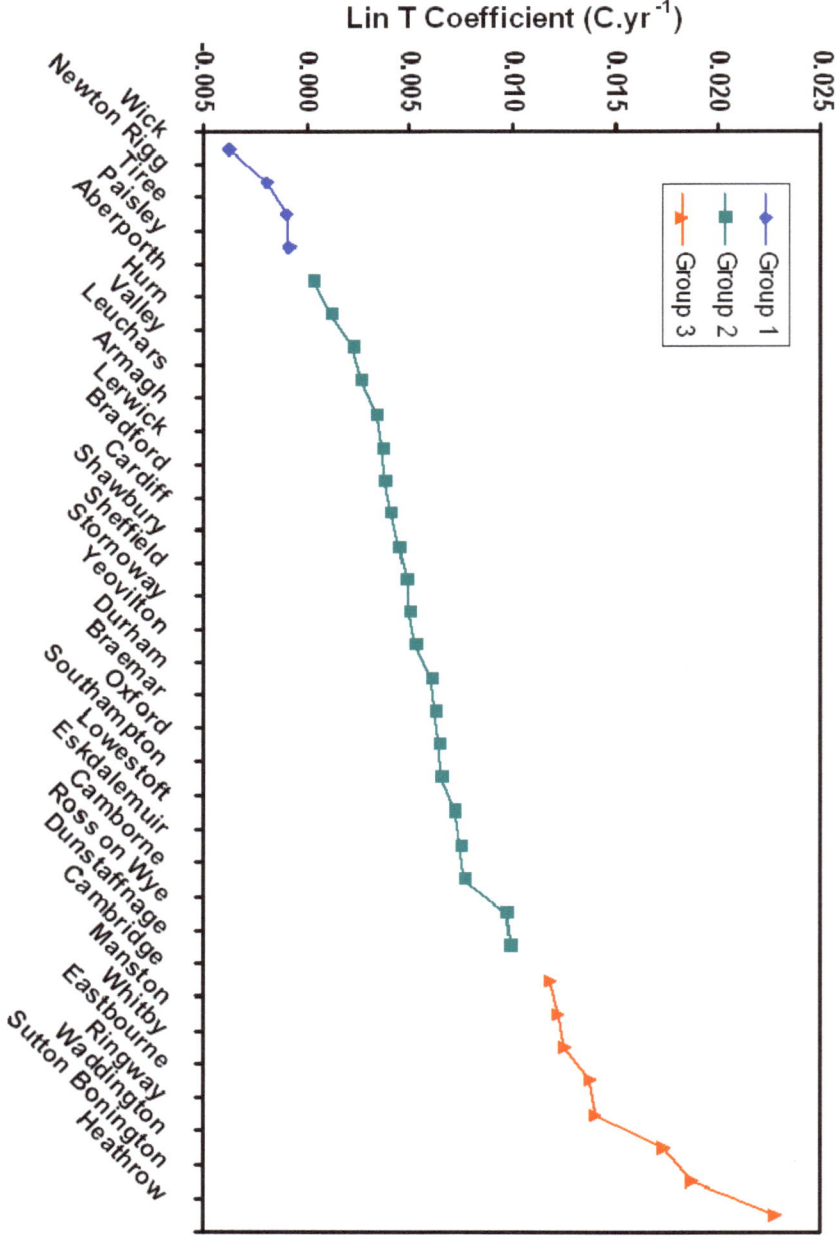

Figure 4-15: Linear warming trend data for the UK weather stations. The stations were divided into three groups based on linear trend magnitude.

It is also interesting to compare the ocean surface temperature trends to the 'predicted' change in 'surface temperature' using the 'radiative forcing constant' for CO_2. This is shown in Figure 4-16. The AMO (green), the PDO (blue) and their linear trends from 1960 are plotted along with the change in 'surface temperature' derived from the increase in downward CO_2 LWIR flux and the 'radiative forcing constant' for CO_2.[14] This is the thick red 'hockey stick' line in the figure. The light orange line is the average of the AMO and PDO trend lines. When this is offset by 0.243 C it overlaps very well with the hockey stick 'prediction'. This should make it clear that the observed change in 'surface temperature' has nothing to do with CO_2, but is related to changes in ocean surface temperatures. While the data processing used to generate the 'hockey stick' has been deliberately concealed, it is reasonable to expect that the combination of urban heat island increases and temperature 'homogenization' or 'adjustment' would account for the observed offset.

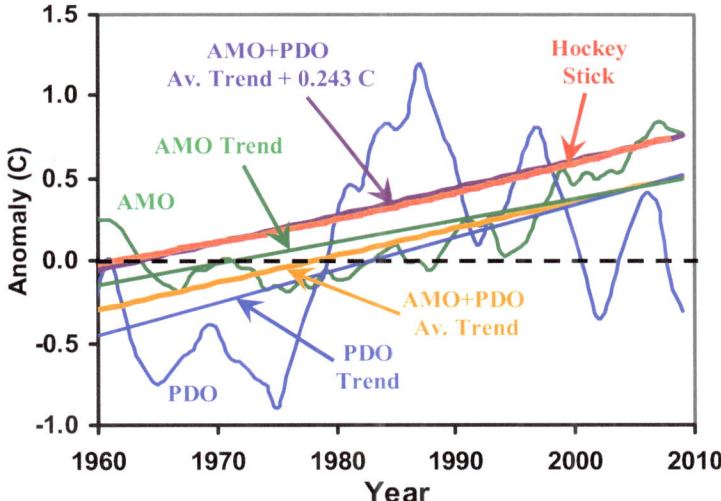

Figure 4-16: AMO and PDO and trend lines plotted from 1960. The hockey stick surface temperature prediction is also shown. When the average AMO+PDO trend line is offset by 0.24 C it almost matches the hockey stick prediction.

4.1.3 The Night Time Air Temperature Transition from Convection

During the day, the ground surface cools by (moist) convection because the solar heated surface is at a higher temperature than the air immediately above it. As the air cools after sunset, a point is reached when the air temperature close to the surface is approximately equal to the surface temperature. Convective cooling stops and the surface continues to cool by LWIR emission through the LWIR atmospheric transmission window. This is illustrated above in Figure 4-4. The ground surface temperature may now cool below the surface air temperature. The air also cools by thermal conduction to the surface. This depends on the local wind speed, near surface turbulence and air drainage flow. Cooler air will sink to the lowest local surface level. In addition, as the air cools, condensation often occurs with the release of latent heat (dew formation). While the details of the cooling process are complex, it is driven by the net loss of heat by LWIR emission from the surface. In addition, as the weather conditions change, the minimum surface air temperature is set by the bulk air temperature. The surface will cool by convection until it reaches the local surface air temperature. Figure 4-17 shows the minimum surface and air temperatures for days 60 to 180 using the data from Figure 4-1. The daily minimum air temperature shows distinct spikes of approximately 10 C caused by the ocean-desert weather shift. The surface temperature clearly follows the air temperature, although the surface temperatures are 8 day satellite averages. The places where the satellite averaging closely follows the air temperature shifts are indicated by the arrows.

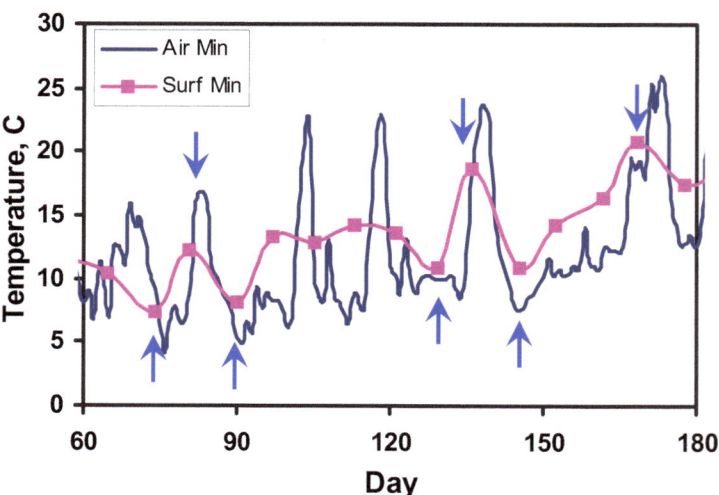

Figure 4-17: Minimum surface and air temperature for days 60 to 180 from Figure 4-1.

When the prevailing weather conditions are from the ocean, the night time temperatures will be linked to the ocean surface temperatures in the region of origin of the weather system. These in turn will influence the ground surface temperatures. The night time transition from convection is therefore an important factor in setting the surface temperature and the local climate. This is physics underlying the PDO 'signature' in the California minimum MSAT weather station data and the AMO 'signature' in the corresponding UK data.

4.2 The Air-Ocean Interface

Only the sun can heat the ocean and cause climate change. This rather obvious fact has been conveniently ignored in the global warming 'debate' even though it follows very simply from a consideration of the energy transfer processes that occur at the Earth's air-ocean interface. The optical transmission of water reaches a maximum in the blue green visible spectral region near 475 nm. This allows sunlight to penetrate and warm the oceans to depths of up to 100 m. The transmission decreases with both increasing and decreasing wavelength as shown above in Figure 3-12. The penetration depth of long wave infrared (LWIR) radiation is less than 100 micron. This is the typical width of a human hair. The oceans are heated by sunlight and cool at the surface through a combination of evaporation and LWIR emission. The cooled surface water sinks and is replaced by warmer water from below. The subsurface layers may be warmed or cooled by the descending surface water, depending on the thermal gradient. Evaporation increases significantly with ocean surface temperature and also increases approximately linearly with wind speed. The long term zonal averages of ocean temperatures and latent heat flux are shown in Figure 4-18.[43] Long term, 50 year average ocean surface temperatures are shown in Figure 4-19.[43] The warm water pools that have a major influence on the Earth's weather patterns can be clearly seen as the dark red regions near the equator. Long term, 50 year average evaporation rates are shown in Figure 4-20.[44] It is important to note that the peak evaporation rates do not coincide with the peak surface temperatures in the warm pools. This is because evaporation depends on both surface temperature and wind speed.

Historically, data on ocean subsurface temperatures has been sparse. However, this has changed recently with the implementation of the Argo Program and other float and buoy measurement programs.[45] Any discussion of climate change must explicitly include ocean solar heating and long range ocean transport since these are the primary mechanisms of climate change. It is simply impossible for the observed 100 ppm increase in atmospheric CO_2 concentration to cause any kind of change in ocean temperature. It is also important to understand that ocean temperatures are transport or kinetically limited by ocean mixing and the ability of the oceans to transfer heat to the surface where cooling can occur. There is no thermal equilibrium on any time scale.

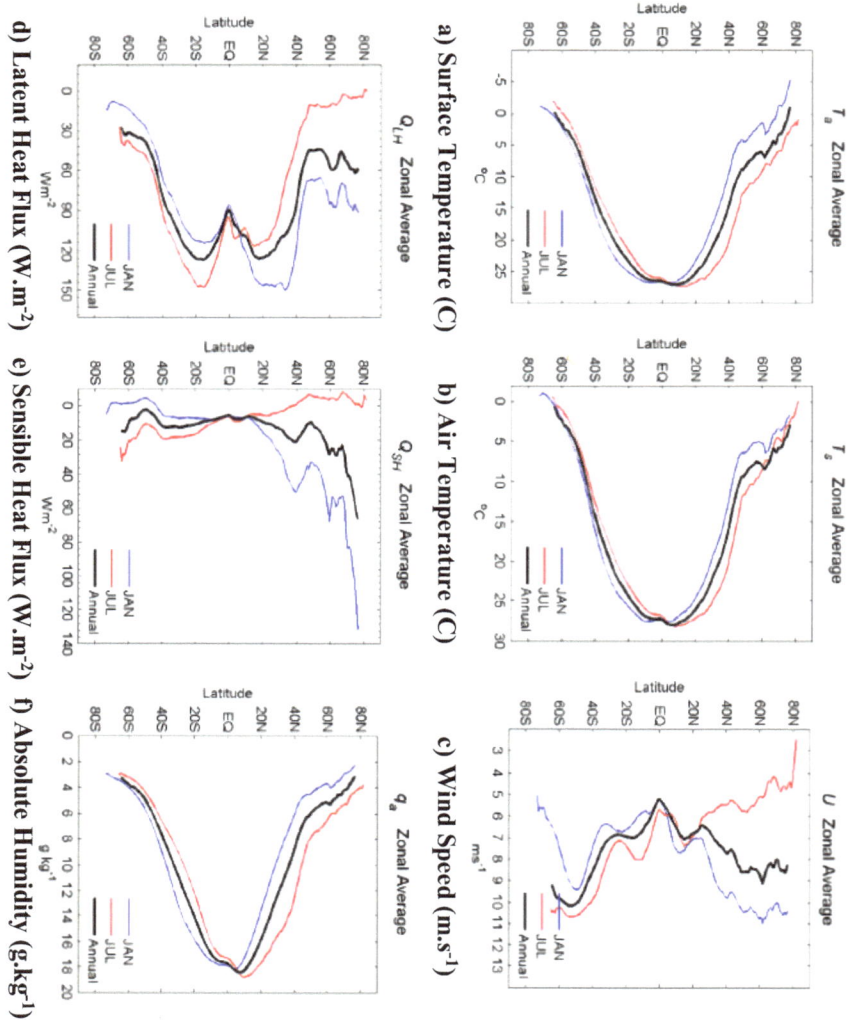

Figure 4-18: Latitude zone dependence of long term (50 year) averages of ocean and air temperatures, wind speed, latent and sensible heat flux and absolute humidity.[43]

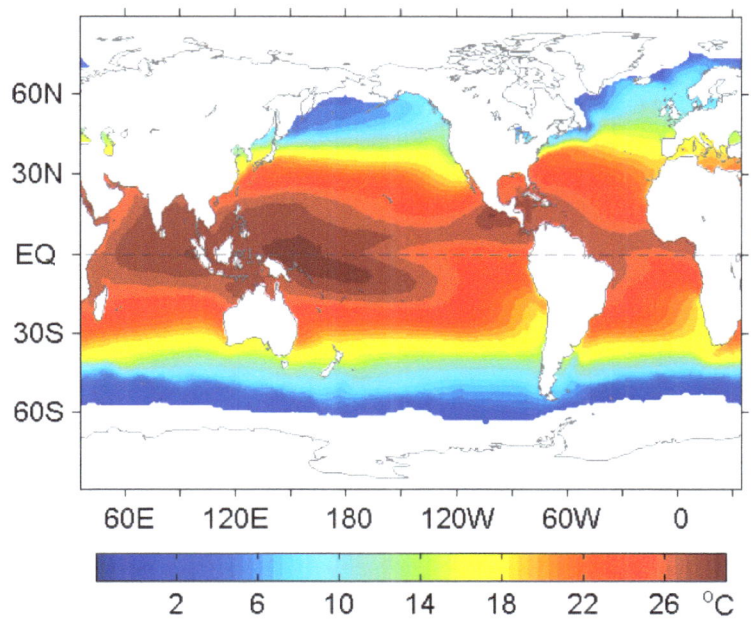

Figure 4-19: Fifty year average annual ocean surface temperatures. [43]

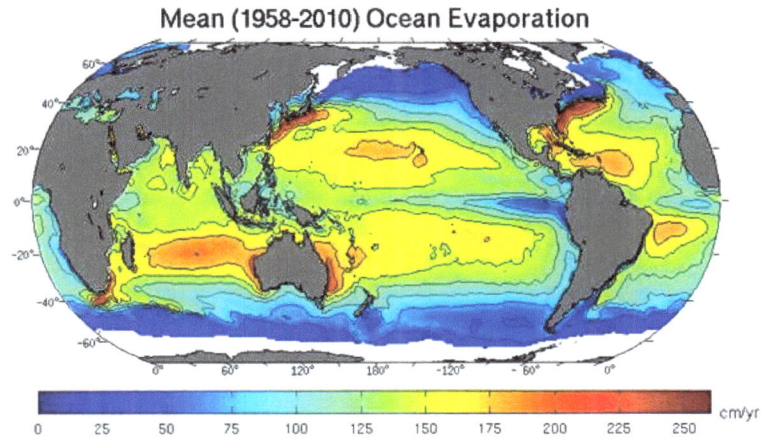

Figure 4-20: Average long term (50 year) ocean evaporation rates. [44]

4.2.1 Ocean Heating From Argo Float Data

The Argo float program started in 2000 and reached full deployment in 2007. High quality subsurface ocean temperature, salinity and density data are now available from a fleet of 3000 submersible floats that are distributed throughout the world's oceans.[45] The floats are designed to sink to a depth of 1000 or 2000 m, drift at that depth for ten days, then return to the surface, acquiring temperature, density and salinity data during the ascent. At the surface, the data are transmitted via satellite to a series of ground monitoring stations and most of the data is published electronically within 24 hours. The floats then repeat the descent/ascent cycle. The floats are not tethered and drift with the ocean currents. The principal features of the solar heating of the ocean at various latitudes through the year may be understood by examining the results from selected Argo floats. The regular annual variation in the ocean temperature profile may be understood by examining the data from a float drifting near the Tropic of Capricorn in the central S. Pacific Ocean. Near the equator, the diurnal mixing depth is not sufficient to mix the ocean layers so heat can accumulate below the mixing layer for extended periods. This can be seen in the data from a float drifting near the equator. Further details and additional float data are provided in Reference 3. Figure 4-21 shows the variation in ocean temperature for an average latitude of 20.9° S in the central Pacific Ocean. The temperatures at 13 depths, down to 155 m are shown as a time series for the year 2007. The latitude and longitude plots of the float drift are also shown. The seasons are reversed in the southern hemisphere, so the peak summer temperatures are reached in March following the peak solar intensity in December. The thermal gradient structure begins to form in November and the thickness of the diurnal mixing layer decreases. As the ocean cools in the fall, the depth of the mixing layer increases. In this region, the ocean is fully mixed down to 150 m for about 3 months from August through October. This clearly demonstrates the seasonal thermal hysteresis in the ocean temperature. Because of variability in the float actuators, the depths are averages with a standard deviation of approximately 0.2 m. As the latitude increases, the temperatures decrease and the time over which the layers are fully mixed also increases.

Near the equator, the ocean temperature profile is rather different. Figure 4-22 shows float data at an average latitude of 1.5° S. The average temperature at 5 m is 24.5 C. There is no obvious seasonal peak in the data. In this example, the temperatures decreased over the year as the float drifted eastwards. Mixing down to the 40 m level occurs most of the time.

However, the 75 and 100 m levels have temperatures that are on average 6 and 8 C below the 40 m level. These lower levels are rarely coupled to the surface. This means that solar heat can accumulate within these subsurface levels for extended periods. This is the origin of the ENSO fluctuations in the equatorial Pacific Ocean. It also explains variations in cyclone intensities. At higher latitudes, the ocean cools and the subsurface layers are coupled to the surface during the winter months. The ocean has to cool to -1.8 C for ice to form.

It is also important to note that at high latitudes, the surface area of a spherical zone decreases significantly. This geometric factor increases the depth of ocean currents as they flow to higher latitudes, further limiting their interaction with the surface. This is shown in Figure 4-23. Small changes in subsurface ocean temperatures can therefore result in large changes in polar ice formation. No interaction with the atmosphere is required and changes in atmospheric CO_2 concentrations can have no effect on this process.

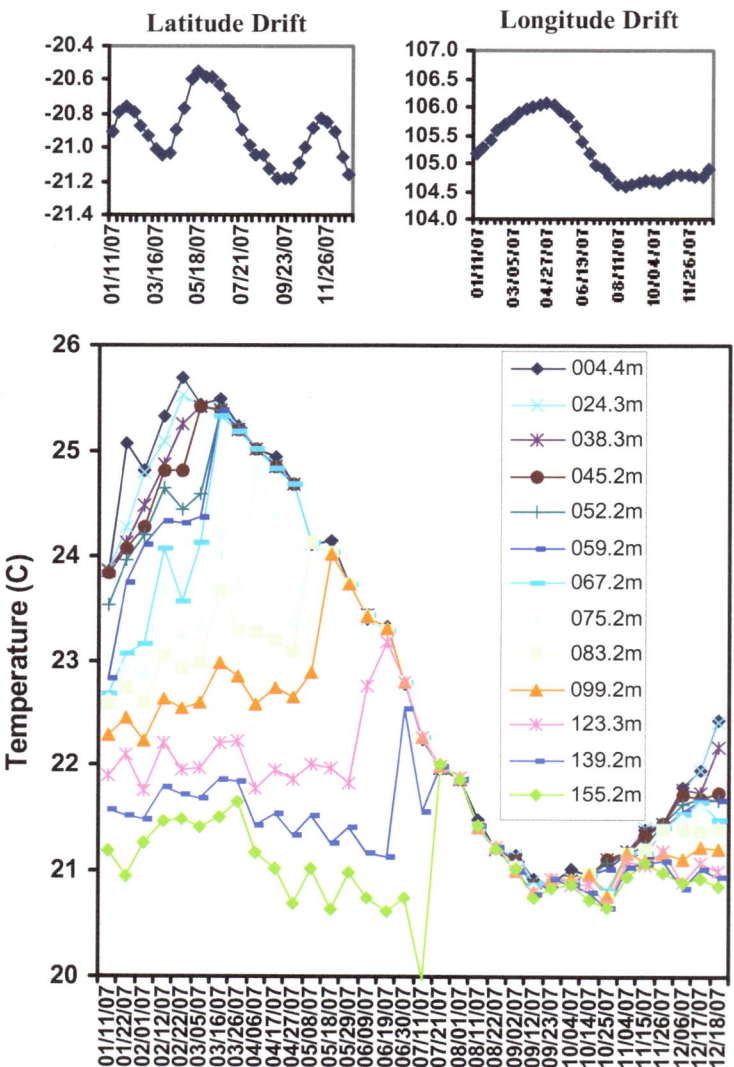

Figure 4-21: Argo Float Data for 20.9° S average latitude.

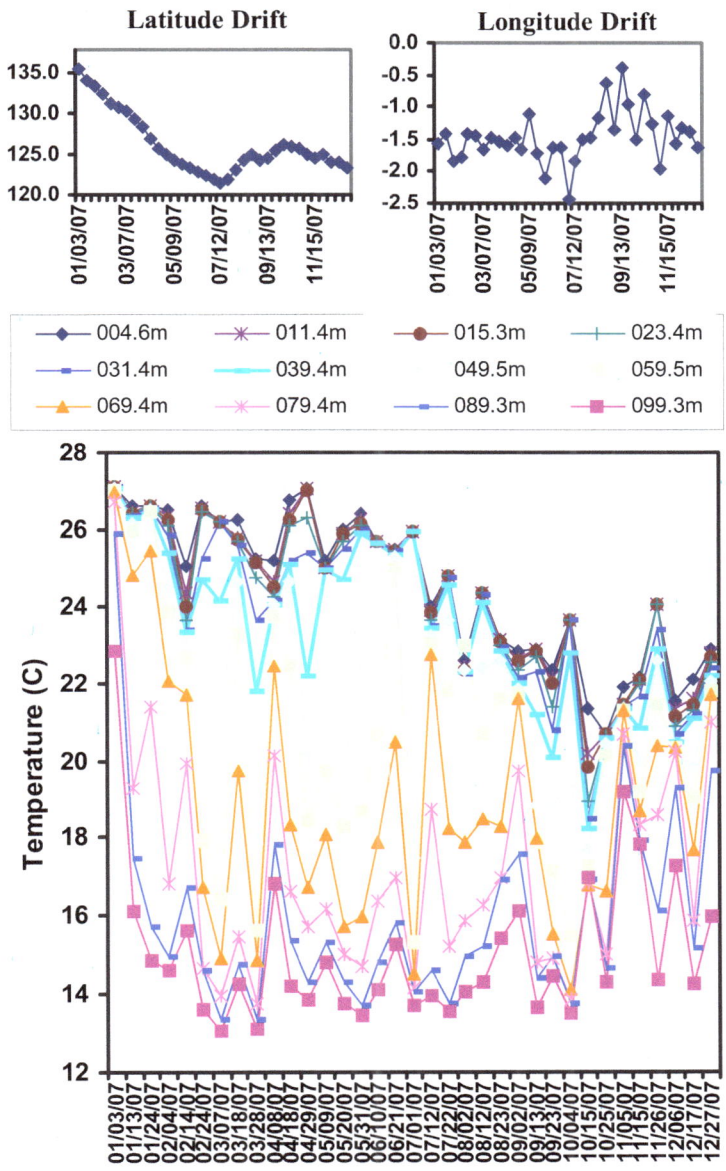

Figure 4-22: Argo Float Data for 1.5° S average latitude.

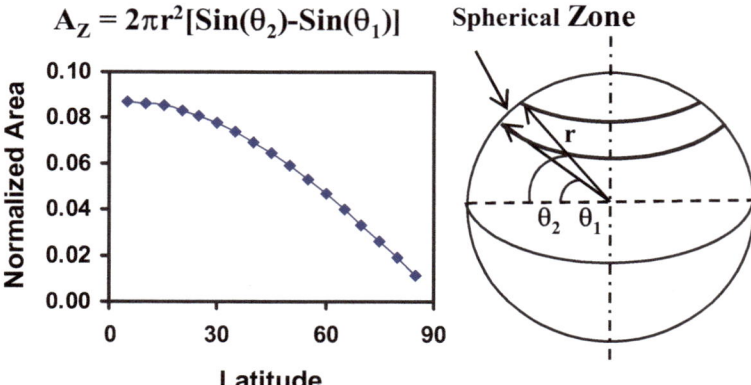

$$A_Z = 2\pi r^2[\text{Sin}(\theta_2)-\text{Sin}(\theta_1)]$$

Figure 4-23: Decrease in the area of a spherical zone with latitude (5° zones).

4.2.2 The Simulation of Ocean Solar Heating

The solar heating of the ocean may be simulated using a simple Beer's law model as illustrated in Figure 4-24.[3] The solar flux after transmission through the atmosphere was calculated as a function of latitude, day of year and time of day using the 'clean air' polynomial adapted from the IEEE 738 standard. This is used to calculate the solar heating and sag of electrical power lines. The surface reflection and transmission at the air ocean interface were determined from the Fresnel equations for the angle of incidence of the solar radiation with the refractive index of water set to 1.333. This was used to determine ocean solar heating and cooling as a function of depth over a 1 year period at 30° latitude. The model depth resolution was 1 m and the time step was 30 minutes. A single scattering coefficient of 0.075 m^{-1} was used together with a single constant cooling flux term. The cooling term was set so that the model was stable and returned to the same surface temperature after each annual cycle. The calculated results are shown in Figure 4-25. They are consistent with Argo Float data such as the example shown in Figure 4-21.

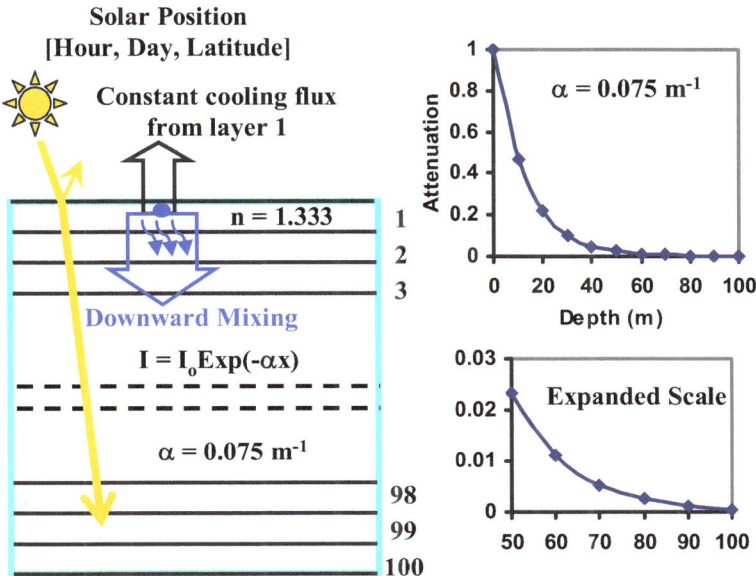

Figure 4-24: Ocean solar heating model based on the solar atmospheric transmission algorithm from IEEE 738.

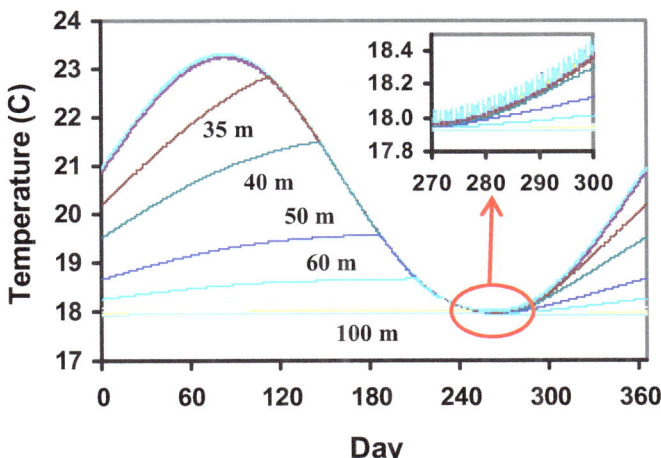

Figure 4-25: Ocean layer solar heating model results. Ocean layer solar heating to 100 m depth calculated over a 1 year period for 30° latitude. As the ocean heats up in the spring and summer, stable layers form. As the ocean cools in the fall and winter, the heated layers cool from the surface down. Inset shows the diurnal mixing layer and the initial heated layer separation.

The model was then extended to simulate the effects of changes in the solar constant due to the sunspot cycle from 1650 to 2009 using a scale factor of 1 $W.m^{-2}$ for a 100 change in the annual sunspot index.[3] The changes in solar flux calculated from the sunspot index are shown in Figure 4-26. The calculated changes in ocean temperatures at 90 m depth are shown in Figure 4-27. There is a distinct decrease to the end of the Maunder Minimum followed by an overall increase of almost 0.5 C from 1750 to 2000. This simple model clearly demonstrates that small changes in the solar constant influence ocean temperatures can cause climate change. Subsurface ocean layers are transported over long distances by wind driven ocean currents without any interaction with the surface. The largest changes in temperature occur at lower latitudes.

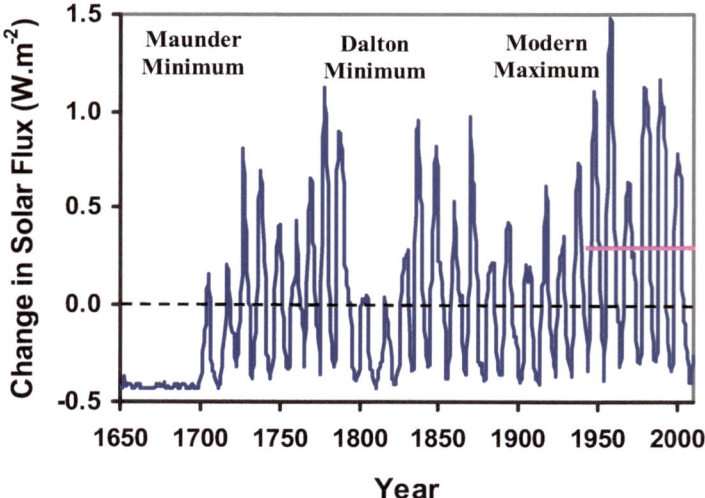

Figure 4-26: Calculated change in the solar flux due to sunspot activity from 1650.

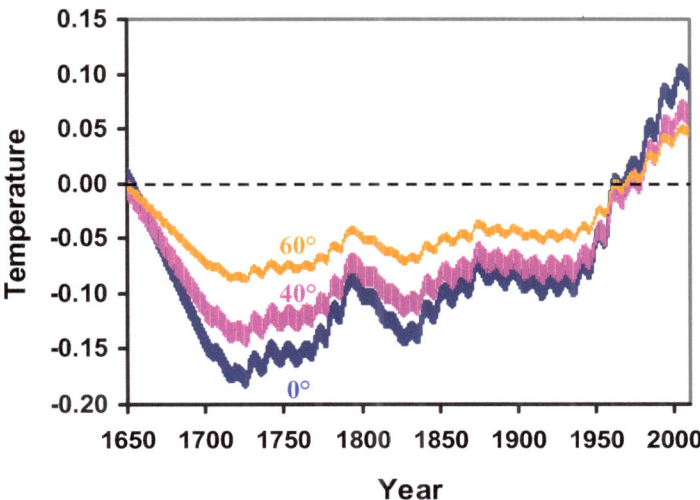

Figure 4-27: Calculated changes in ocean temperature from 1650 at 90 m depth and 0, 40 and 60° latitude.

The measured average global ocean temperature increase for the 0 to 300 m depth level from 1953 to 2003 was 0.17 C and the average increase in flux needed to heat the oceans was 0.2 W.m^{-2}.[46] This is consistent with Figure 4-27 over the same time period. However, there was also significant variation in the measured temperatures between ocean basins. This is shown in Figure 4-28. The N. Atlantic warmed by 0.35 C, the N. Pacific by 0.09 C. These fluctuations are caused by differences in ocean circulation, mixing and wind speed. For example, part of the southern equatorial current in the Atlantic Ocean is diverted northwards by the coast of Brazil.

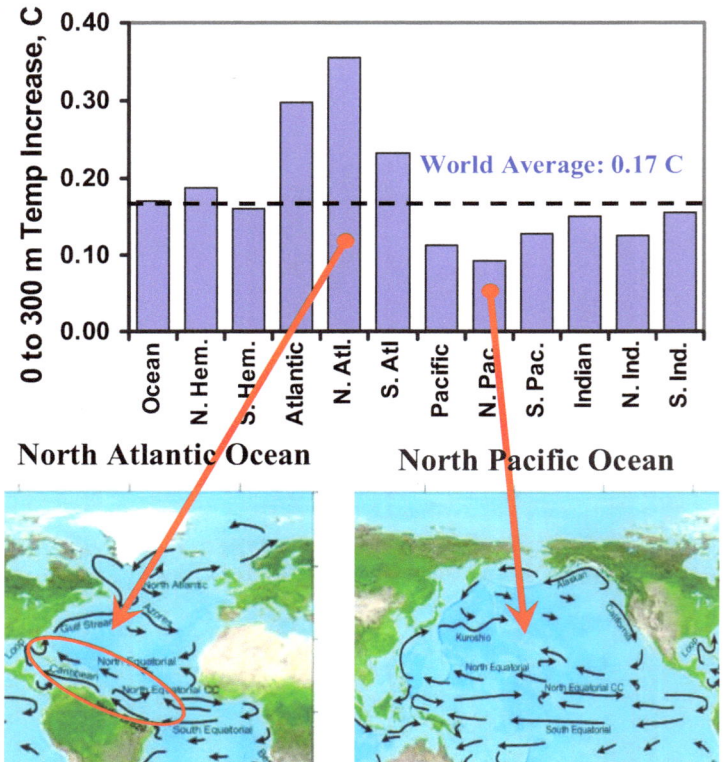

Figure 4-28: Changes in ocean temperatures, 0 to 300 m from 1955 to 2003.[46]

It is also straightforward to show that a 1.7 W.m^{-2} increase in downward atmospheric LWIR radiation from a 100 ppm increase in CO_2 cannot cause ocean warming. As shown above in Figure 3-12, the penetration depth of LWIR radiation into the ocean is less than 100 μm. This is about the width of an average human hair. The surface interaction volume over a 1 m^2 area is very small, closer to 1 cm^3 than 10 cm^3. A small change LWIR radiation coupled to the surface interaction volume will cause a rapid change in surface temperature which in turn will change the surface evaporation rate. A flux of 1.7 W.m^{-2} represents a water evaporation rate of 2.7 g.hr^{-1}m^{-2} for ideal 'clear sky' conditions. This corresponds to a 2.4 cm.yr^{-1} increase in evaporation rate since 1800, with 1.7 cm.yr^{-1} of this increase occurring over the last 50 years. Global estimates of ocean evaporation rates show that between 1977 and 2003 the global ocean evaporation rate has increased from 103 to 114 cm.yr^{-1} with an uncertainty of ±2.72 cm.yr^{-1}.[47] This was caused by a 0.1 m.s^{-1} increase in average wind speed. The 'clear sky' upper limit for the CO_2 induced increase in evaporation is below the measurement uncertainty bounds. The evaporation rates are shown in Figure 4-29.

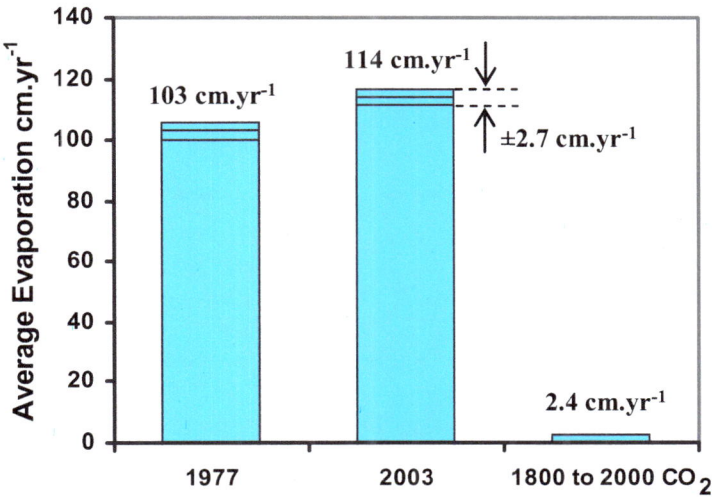

Figure 4-29: Changes in ocean evaporation rates from 1977 to 2003 and the 'clear sky' upper limit to the evaporation rate increase due to a 1.7 W.m^{-2} increase in LWIR flux from CO_2. [47]

Long term averages of surface air temperatures are approximately 2 C below the corresponding ocean surface temperatures. This means that there is usually no direct heating of the ocean by the atmosphere, as required by the Second Law of Thermodynamics. Latent heat of evaporation is not released until the water condenses, which is generally at altitudes above 1 km. It is therefore impossible for an increase in downward atmospheric LWIR flux of 1.7 W.m^{-2} to heat the ocean. The increase in flux is converted at the ocean surface into an insignificant change in evaporation rate. This is buried in the noise of wind induced fluctuations in evaporation and changes in LWIR flux caused by variations in solar illumination, aerosols, cloud cover and near surface humidity.

4.2.3 The Pacific Warm Pool

As the water is transported across the Pacific Ocean by the N. and S. Pacific Equatorial Currents, it is warmed by solar heating. This warm ocean water accumulates in the Pacific warm pool in the western Pacific Ocean. This can be seen above in Figure 4-19. The heating of the Pacific warm pool may be understood by examining the ocean temperatures recorded by the TAO/TRITON Buoy network. This monitoring network was established to monitor the ENSO variations in the tropical Pacific

Ocean. The location of the buoys is shown in Figure 4-30. Buoy data are available from 1979. The ocean parameters recorded vary with time and location. Blocks of data may be missing because of sensor failure.

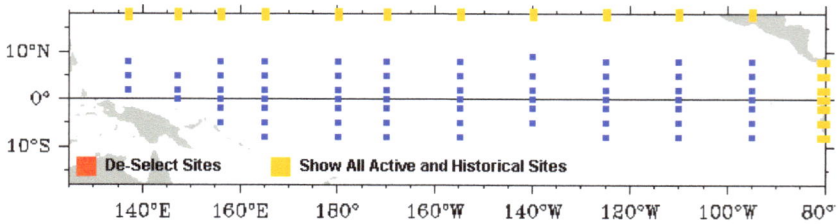

Figure 4-30: TAO/TRITON network buoy locations.

Figure 4-31 shows the change in average surface temperature along the equatorial TAO/TRITON buoys for the January 2005 to February 2011 period. The average daily solar flux is also shown. The error bars are the one sigma points. The surface temperature increases linearly from 23.8 C at 110° W to 29.6 C at 156° E (204° W). The temperatures then level out and the average surface temperature does not exceed 30 C. The last point is from 137° E, 2° N because there is no buoy on the equator at this longitude. The daily solar flux reaches a peak of 24.6 $MJ.m^{-2}.day^{-1}$ at 140° W. The daily flux then decreases to 18.6 $MJ.m^{-2}.day^{-1}$ and the standard deviation increases from ±3 to ±6 $MJ.m^{-2}.day^{-1}$ moving westward along the buoy network. The decrease in solar flux is consistent with an increase in cloud cover.

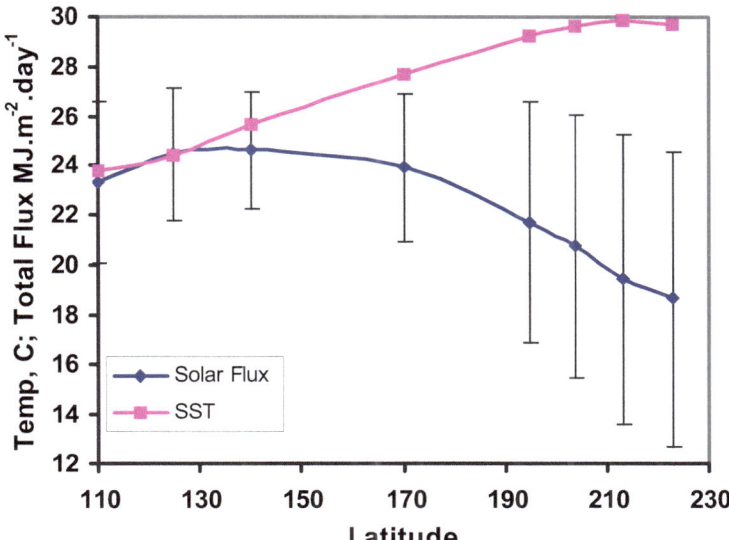

Figure 4-31: Variations in average SST and solar flux along the TAO/TRITON equatorial buoys, January 2005 to February 2011. The error bars are the one sigma standard deviation points.

Figure 4-32 shows the changes in average ocean temperatures from January 2005 to February 2011 in the equatorial TAO/TRITON buoy data down to depths of 200m. The increase in temperature from 110° W to 165° E at 150 m is 9 C and at 200 m it is 4 C. Figure 4-33 shows the corresponding depth resolved change in the ocean heat content in $MJ.m^{-3}$ from 110° W to 165° E derived from a linear interpolation of the temperature data in Figure 4-32. The heat content increases by 25 $MJ.m^{-3}$ near the surface and by over 50 $MJ.m^{-3}$ near 100 m depth. The heat content increase near 200 m depth is still over 25 $MJ.m^{-3}$. By comparison, the peak daily solar flux is ~25 $MJ.m^{-2}.day^{-1}$. Figure 4-34 shows the accumulation of ocean heat between the individual buoys. The accumulation shifts to lower depths with increasing longitude from W to E. From 110° to 125° W, the peak heat accumulation is ~16.5 $MJ.m^{-3}$ between 60 and 80 m. From 155° to 170° W, the peak heat accumulation is ~12.5 $MJ.m^{-3}$ between 120 and 140 m. Figure 4-35 shows the raw daily SST data used in the temperature averaging for Figure 4-31. The eastern buoy at 110° W records the lowest temperatures and the largest temperature fluctuations with time. These fluctuations are from the Humboldt Current off the coast of S. America as it turns across the E. Pacific to start the S. Equatorial Current portion of the S. Pacific gyre. The temperature fluctuations are due to the solar heating of the S. Pacific Ocean and gyre related transport and mixing. As the ocean water is transported across the Pacific Ocean, it is warmed by solar heating. The temperatures increase and the temperature fluctuations decrease. An important point that does not seem to be well understood is that the heating is part of a flow system. It depends upon the flow velocity as well as the wind speed along the equatorial current. These two work together to drive the solar heating. Inspection of Figure 4-35 reveals a time delay between the solar signals at 110° and 170° W of approximately 4 months. Because of short term fluctuations and missing data blocks, it is difficult to determine accurate arrival times between the buoys. At 170° W, there was no time delay in the arrival time of the temperature signals at different depths. The whole body of water down to 100 m was moving at the same velocity because of turbulent and diurnal mixing.

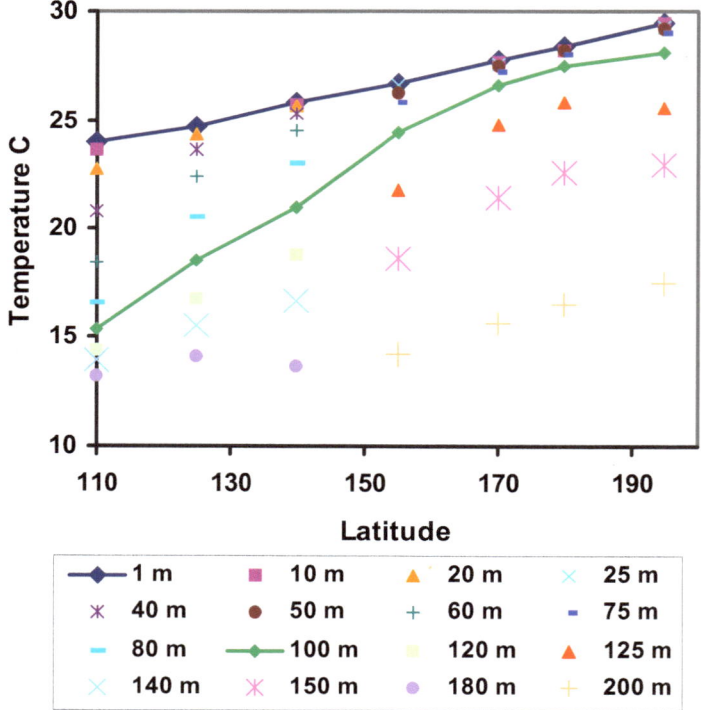

Figure 4-32: Increase in average ocean temperature along the equatorial TAO/TRITON buoys, Jan. 2005 to Feb 2011.

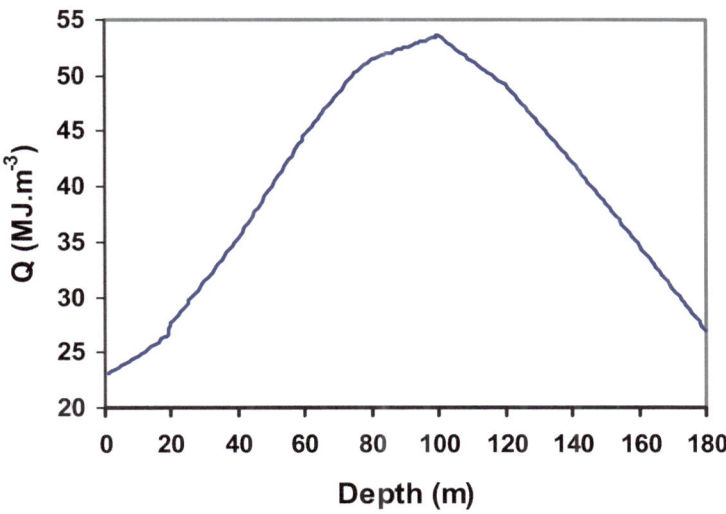

Figure 4-33: Change in equatorial ocean heat content (MJ.m^{-3}) vs. depth from 110° W to 165° E derived from the temperature data in Figure 4-32.

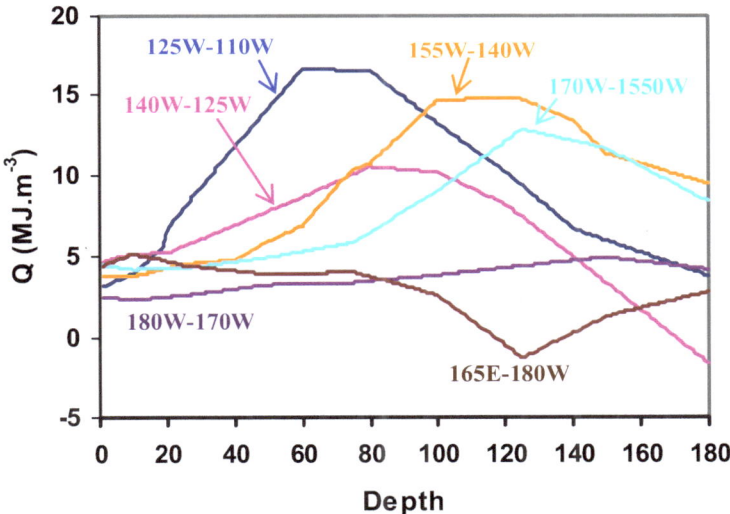

Figure 4-34: Change in heat content between TAO/TRITON equatorial buoy stations (MJ.m^{-3}) vs. depth derived from Figure 4-32.

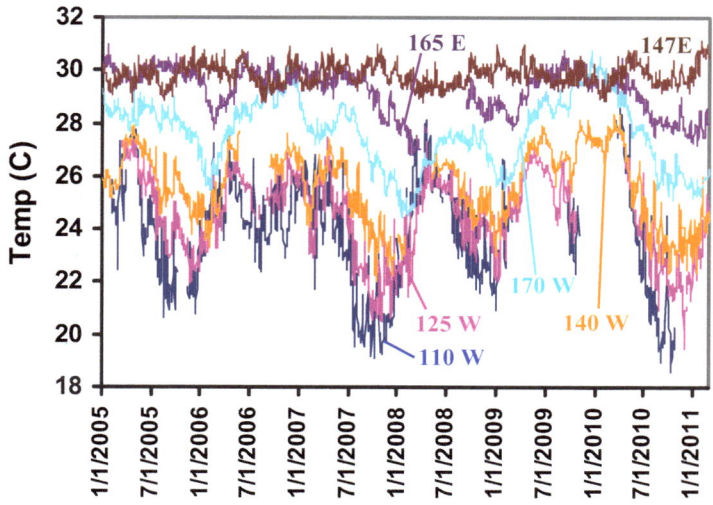

Figure 4-35: Raw TAO/TRITON SST data used in the temperature averaging.

4.2.4 Ocean Cooling and the Energy Balance at the Air-Tropical Ocean Interface

The 'clear sky' tropical solar flux is approximately 25 MJ.m^{-2}.day^{-1}. Hourly data are available from the TAO/TRITON buoy at 156° E located on the equator. These data include the solar flux, the wind speed, the relative humidity, precipitation, the air temperature and ocean temperatures at depths including 1.5, 25 and 50 m. Using these data, a simple model of the ocean diurnal heating cycle was constructed. The solar flux was used as measured. The latent heat flux was calculated from the SST, the air temperature and the relative humidity using an adjustable fitting parameter corresponding to the turbulent exchange coefficient.[47] The ocean was divided into 0.5 m layers and the total absorption fraction was used at each depth to calculate the solar heating. The absorption fractions used are shown above in Figure 3-13. Nearly half of the solar flux was absorbed in the first 0.5 m layer. The cooling flux was coupled into the first 0.5 m layer. In addition to the wind driven evaporation, a fixed LWIR cooling flux was added to simulate the emission through the LWIR window. A small convection coefficient was also included. The surface cooling and the solar heating terms were calculated to a depth of 30 m. The layers were then mixed based on temperature using a 5 layer exponential diffusion term to simulate the surface mixing. The model parameters were adjusted to fit the measured temperature data at 1.5 and 25 m depths.

Hourly TAO/TRITON buoy data for 0° lat, 156° E were downloaded and used 'as is'. These data are plotted in Figure 4-36 for July to December 2010. The first 40 days of data are plotted on an enlarged scale in Figure 4-37a) and b). There is a clear inverse relationship between the diurnal peak in the ocean temperature at 1.5 m depth and the wind speed. The diurnal temperature peaks increased when the wind speed was less than ~4 m.s^{-1}. This can be seen in areas indicated by the dotted lines. In addition there was a distinct phase shift or time delay between the peak solar flux and peak ocean temperature at 1.5 m depth. This is shown in Figure 3-38 for days 4 to 10. The time delay increases with the increase in the temperature peak.

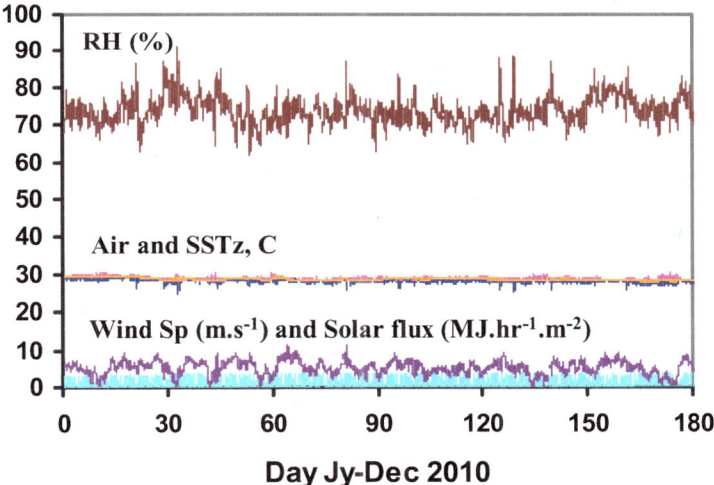

Figure 4-36: Hourly TAO/TRITON data for 0, 156°, July to December 2010.

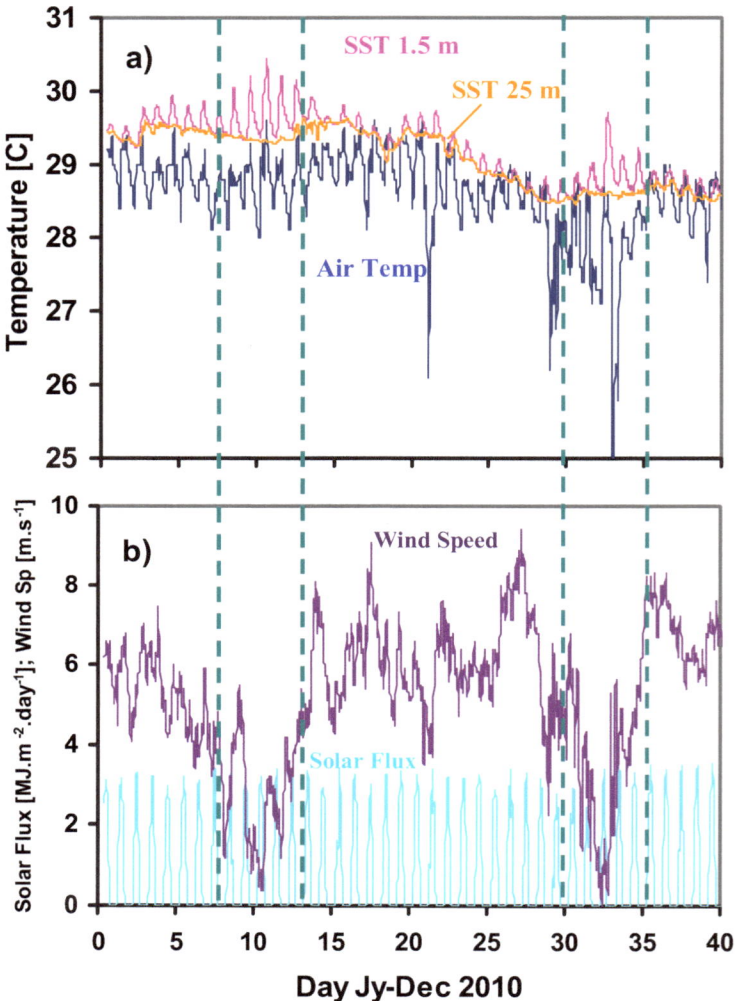

Figure 4-37: a) Air and ocean temperatures, 1.5 and 25 m depths and b) solar flux and wind speed, expanded scale from Figure 4-36.

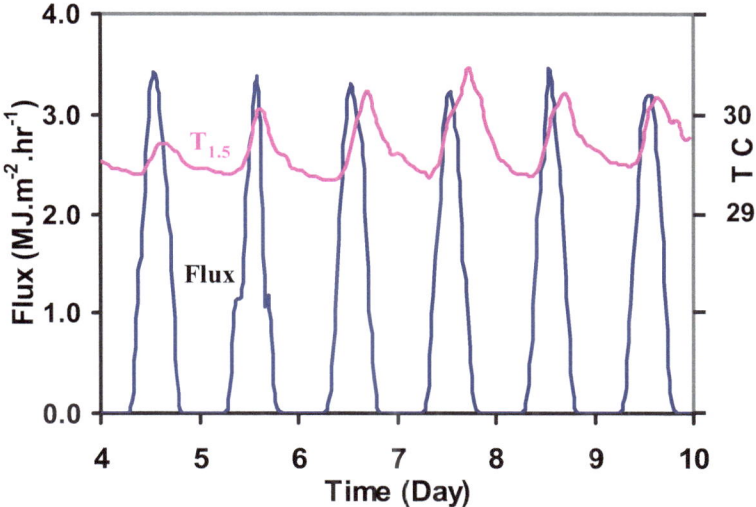

Figure 4-38: Solar flux and 1.5 m ocean temperature for days 5 to 10, showing the time delay between the flux and the temperature peaks.

Figure 4-39 shows some initial model output which simulates both the diurnal surface temperature fluctuations and the underlying temperature trend quite closely. In this case, a quadratic fit to the wind speed was used. The fit coefficients were $3.6U$ and $0.25U^2$. The LWIR transmission window flux was 50 W.m^{-2} and the dry convection coefficient was 5 W.m^{-2}.C^{-1}. The differences between the measured and calculated data are most probably due to ocean transport. The buoy is stationary and the ocean water flows past. The model simulates the same body of water as it is moving in the ocean. The total daily cooling and solar fluxes and the daily average wind speed are plotted in Figure 4-40. The influence of the wind speed can clearly be seen in the cooling flux variations. The cooling flux exceeds the solar flux at the higher wind speeds. However, during the lower wind speed periods as indicated in Figure 4-37, the cooling flux is less than the solar flux. These changes in flux can reach ±10 MJ.m^{-2}.day^{-1}. The total increase in downward atmospheric LWIR flux from an increase of 100 ppm in atmospheric CO_2 concentration is only 0.15 MJ.m^{-2}.day^{-1}. It is simply impossible for the observed increase in atmospheric CO_2 concentration to have any effect on ocean temperatures once the role of the winds speed is understood.

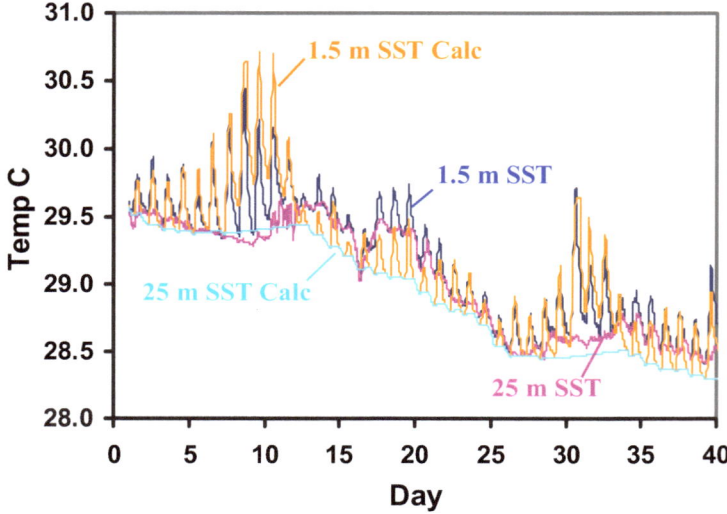

Figure 4-39: Initial model results compared to measured 1.5 and 25 m ocean temperatures.

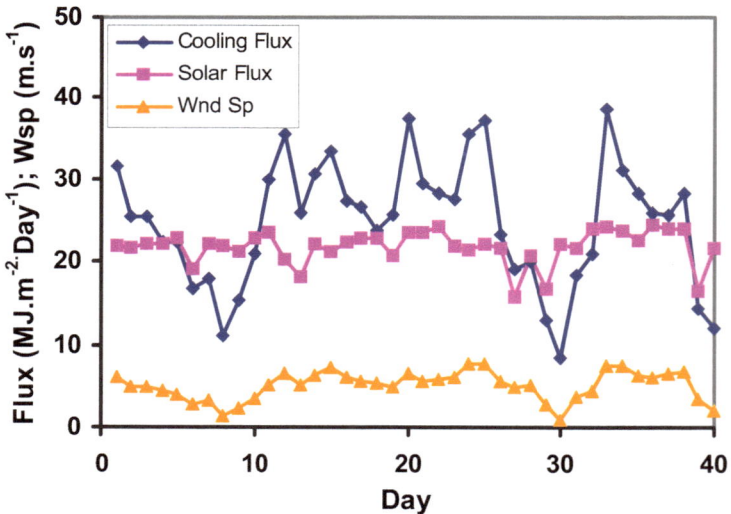

Figure 4-40: Total daily solar flux and model cooling flux showing the strong influence of the wind speed.

4.2.5 Ocean Cooling at Mid Latitudes

At mid latitudes, the ocean is heated by the sun during the summer and cools in winter. A stable thermal gradient forms as the sun heats the ocean and this is removed in winter as the excess cooling increases the depth of diurnal mixing layer. In addition, within the ocean gyres, water from the ocean warm pools is circulated and cooled at these latitudes. The seasonal heat storage is substantial and plays a major role in stabilizing the mid latitude climate. Figure 4-41 shows the cumulative daily solar flux in $MJ.m^{-2}.day^{-1}$ for selected latitudes, calculated using the IEEE 738 'clean air' solar flux algorithm. The summer peaks are near 25 $MJ.m^{-2}.day^{-1}$, but the winter fluxes decrease significantly. At 45° latitude, the winter flux decreases to 5 $MJ.m^{-2}.day^{-1}$. Figure 4-42 shows the ocean temperature profiles for 2007 recorded by an Argo float drifting near 42° S, 153.5° W in the mid S. Pacific Ocean. The peak summer/fall temperatures are between 15 and 16 C at 5 and 25 m depth. During the winter/spring, the ocean cools to 10 C down to depths of 100 m. (Seasons are reversed in the S. Hemisphere). Figure 4-43 shows the change in heat content (deviation from the average) calculated from the temperature data in Figure 4-42. The heat content fluctuates because the float is drifting. However, the total annual change in the heat content is approximately 1000 $MJ.m^{-2}$. Most of this is stored in the first 75m. This heat storage corresponds to ~40 days of full summer sun. The accumulation of heat during the summer and subsequent release in winter stabilizes the Earth's mid latitude climate. It is not just the thermal transport from the Gulf Stream and Japan Current that has a warming effect at mid latitudes. The total annual increase in downward atmospheric LWIR flux for a 100 ppm increase in atmospheric CO_2 concentration is only 50 $MJ.m^{-2}$ and none of this can penetrate more than 100 micron into the ocean surface. It is simply impossible for the observed 100 ppm increase in CO_2 concentration to have any effect on climate.

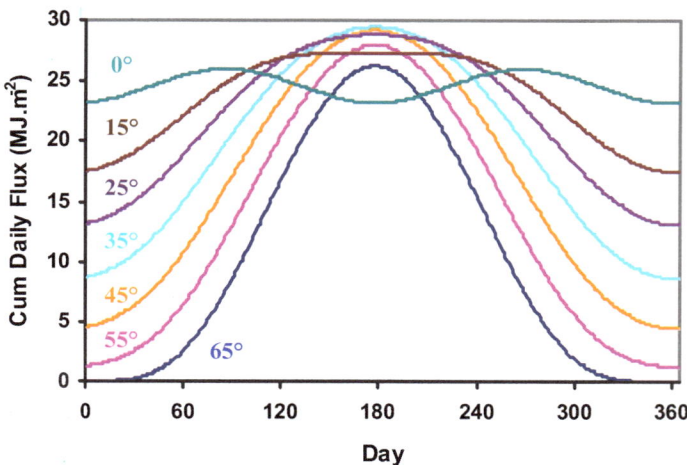

Figure 4-41: Cumulative 'clear sky' daily flux (MJ.m⁻².day⁻¹) at selected latitudes.

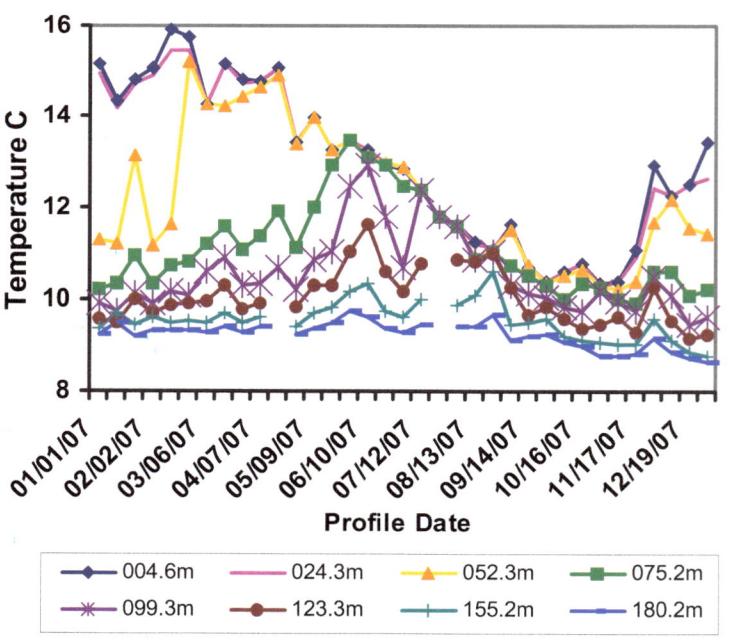

Figure 4-42: Argo float ocean temperature profiles near 42° S, 153.5° W in the mid S. Pacific ocean.

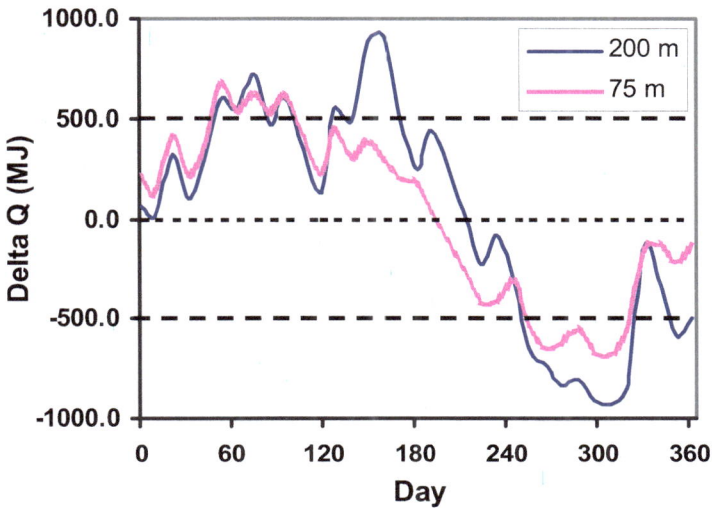

Figure 4-43: Change in ocean heat content calculated from the data in Figure 4-42, MJ.m^{-2} for 75 and 200 m depth columns.

4.3 The Direct LWIR Emission to Space from the Earth's surface

At night, the ground and the air temperatures are usually similar. The dominant energy transfer process is surface cooling by LWIR emission through the atmospheric LWIR transmission window in the 800 to 1200 cm^{-1} spectral region. The magnitude of this flux is typically ~40 W.m^{-2}. Under low humidity conditions it may increase to ~100 W.m^{-2} and downward LWIR emission from low clouds may reduce the cooling flux to zero. The nightly average net LWIR flux for the UCI grasslands site for 2008 is shown in Figure 4-44. The effect of clouds and low humidity can clearly be seen. As the surface and air temperature increase, both the upward and downward LWIR fluxes also increase. Therefore, the direct LWIR emission to space is not a sensitive function of the surface temperature. During the day, when the surface temperature exceeds the air temperature, the flux through the LWIR window increases. However, the direct LWIR emission to space is usually a relatively small fraction of the total daytime surface cooling flux that dissipates the incident solar radiation. Most of the daytime surface cooling is through moist convection. From Figure 4-4 above, under full summer sun conditions in the middle of the day, the net LWIR emission is ~200 W.m^{-2} and the convection is ~800 W.m^{-2}.

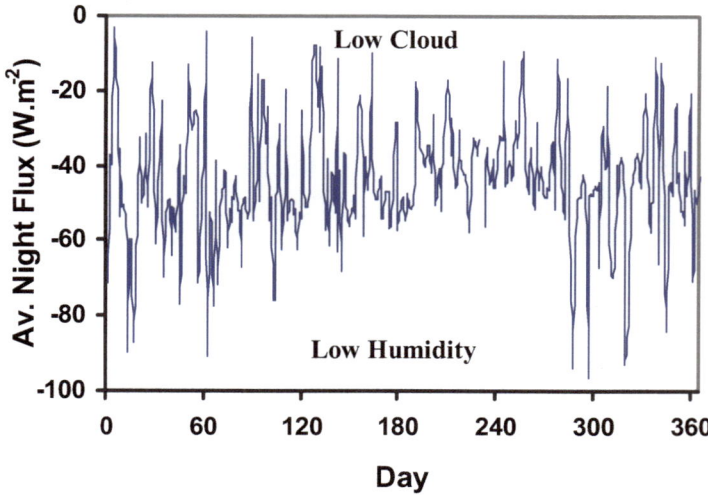

Figure 4-44: Average night time LWIR surface cooling flux for 2008 measured at the UCI 'Grasslands' site.[37] Normal values are in the -30 to -60 W.m⁻² range. The magnitude of the cooling flux decreases with increasing cloud cover and increases under low humidity conditions.

Figure 4-45 shows the calculated normalized downward atmospheric emission at the surface for an air temperature of 300 K and a surface temperature of 302 K. The data are normalized to the surface temperature flux. Only H_2O and CO_2 emission is presented calculated using the truncated data set as described in Reference 3. The normalized blackbody curves for 292 and 277 K are also shown. For a typical lapse rate of -6.5 $K.km^{-1}$ they represent the downward LWIR emission of cloud bases at altitudes of approximately 1 and 4 km. At 292 K, ~85% of the upward emission from the surface through the LWIR window is blocked. At 277 K this decreases to 65%. The typical water content for cumulus cloud is 0.3 $g.m^{-3}$. The total thermal energy needed to evaporate a 1 km cloud column with an area of 1 m^2 will not generally exceed 1MJ.

The effect of clouds on the LWIR emission from the top of the atmosphere depends on the cloud top temperature and the molecular LWIR emission profile above the cloud top. The LWIR emission depends on the temperature profile that sets the water vapor concentration and the line narrowing. The cloud also blocks the upward LWIR emission from the surface in the LWIR window and replaces it with the cloud top emission. While clouds can be readily identified by their visible light scattering properties, the effect of clouds on LWIR emission may not be so obvious. In particular, the differences between the top of atmosphere total clear sky LWIR emission and the total cloudy sky LWIR emission may be small.

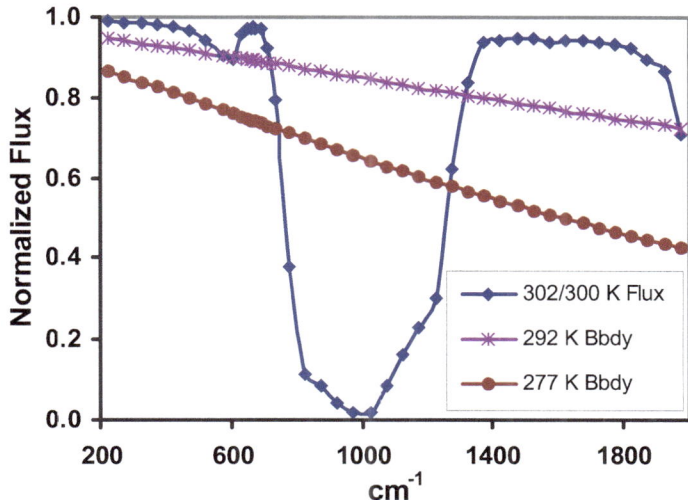

Figure 4-45: Normalized downward LWIR emission, 302/300 K surface/air temperature plotted with 292 and 277 K blackbody curves. For a lapse rate of -6.5 K.km^{-1} these two curves represent the LWIR emission of cloud bases at approximately 1 and 4 km altitude. (Truncated data set, H$_2$O and CO$_2$ only).

4.4 Energy Transfer between the Surface and the Lower Troposphere

Within the atmospheric absorption-emission bands, the downward LWIR flux from the atmosphere balances most of the upward surface LWIR emission and slows the radiative cooling of the surface. The LWIR surface emission is only transmitted through the LWIR atmospheric transmission window. The lower troposphere acts as a 'thermal blanket'. However, there is no thermal equilibrium of any kind. Unless the humidity is low, over 90% of the LWIR flux reaching the surface from the atmosphere originates from within the first kilometer layer of the atmosphere. This is illustrated in Figure 4-46. Similarly, over 90% of the upward LWIR emission from the surface is absorbed by the first kilometer layer. This layer is heated by the surface convection, which resets the air temperature on a daily basis. Excess LWIR emission from the warmer surface is absorbed by the air and produces additional convection. The spectrally resolved absorption of the surface emission is shown in Figure 4-47. However, this absorption only accounts for about 20% of the surface cooling. The air temperature is set by convection, not by 'greenhouse gas absorption'. At night, when the convection slows, the air in the first kilometer layer cools by net upward LWIR emission through the lower troposphere. However, this is a relatively slow process compared to the daytime convective mixing. This allows the lower troposphere to act as a night time thermal reservoir. The air temperature also depends on the local weather system.[35,36]

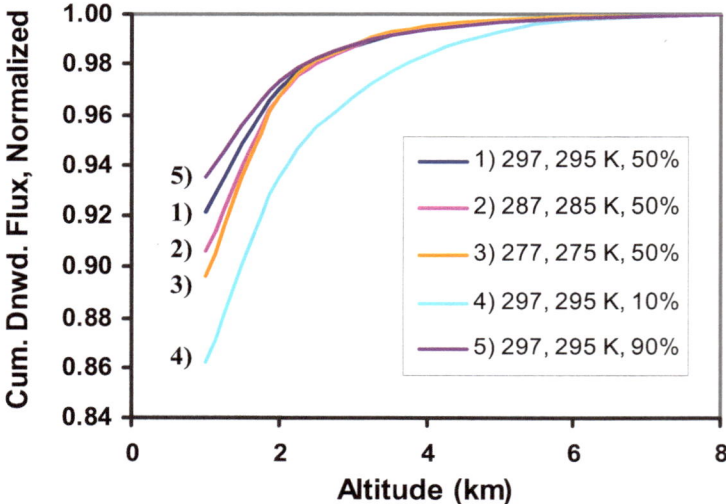

Figure 4-46: Cumulative normalized downward LWIR atmospheric flux vs. altitude, 200 to 2000 cm^{-1}. Most of the flux originates from within the first km air layer. Results from radiative transfer calculations, truncated data set. See Reference 3 for further discussion.

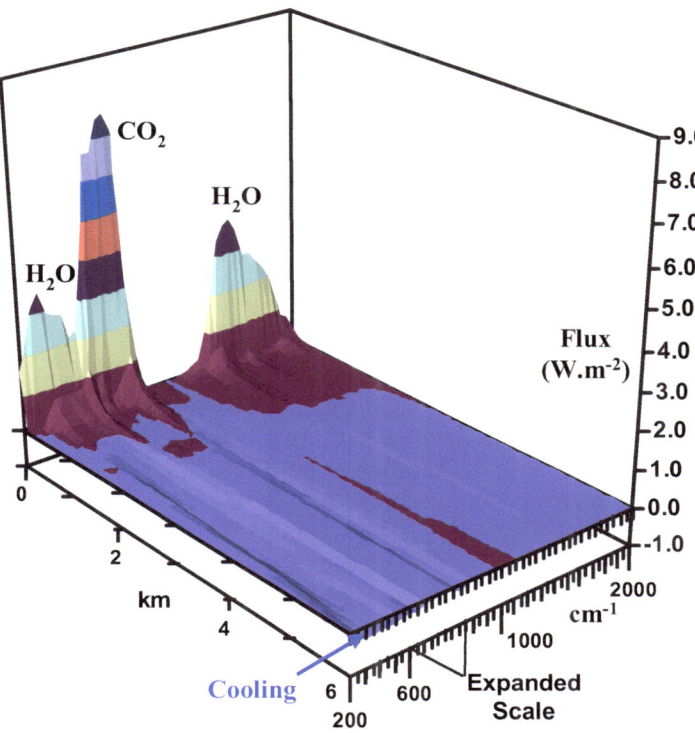

Figure 4-47: Spectrally resolved absorbed flux vs. altitude. Only the H_2O and CO_2 bands are shown. The surface temperature is 325 K, the surface air temperature is 295 K, and the RH is 50%. Most of the LWIR radiation from the ground is absorbed in the first 100 m. (H_2O and CO_2 only - truncated data set). See Reference 3 for further discussion.

4.5 Energy transfer to the Middle and Upper Troposphere

As the daily convective pulse rises through the atmosphere it cools and the water vapor condenses to form clouds above the saturation level. If the air aloft has a low relative humidity, mixing may not result in cloud formation, but in any case the water vapor concentration decreases significantly with altitude as the pressure and temperature decrease. This is illustrated above in Figure 3-10. In the middle to upper troposphere, the daily temperature increase from this convective pulse is generally less than 1 C.[48] There is also a slight increase in the height of the tropopause. At these altitudes, there is a continuous cooling through LWIR emission to space. The peak temperature rise usually occurs during the afternoon. The atmosphere then cools until the next convective pulse arrives. The atmospheric temperature is set by a dynamic balance between convective

89

heating and radiative cooling. There is no equilibrium and the heating and cooling processes are independent. One does not control the other. There can be no radiative forcing.

4.5.1 Latent Heat Release and the Water Vapor Concentration Profile

Large scale climate simulation models usually assume that the relative humidity (RH) remains constant as the temperature changes. This is based on the use of the Clapeyron-Clausius equation to calculate the atmospheric water vapor profile. However, this approach assumes equilibrium conditions and ignores the energy transfer processes that occur during the convective ascent. Observations over the last 60 years have shown a decrease in specific humidity at 7 km altitude from 0.48 to 0.42 g/kg.[49] This means that the climate models are incorrect and that their assumptions of 'water vapor feedback' have no basis in physical reality.[50]

As an air parcel rises through the atmosphere, there is little mixing with the surrounding air and the ascent is assumed to be adiabatic. Energy is conserved within the air parcel. Any changes in temperature within the air parcel must come from a redistribution of the internal energy. The release of latent heat warms the air parcel and causes it to expand and rise further through the atmosphere. This may be expresses formally as follows:

$$dU = C_v dT + Ldq + gdh - PdV = 0 \qquad (4.2)$$

Here, U is the internal energy, Cv is the specific heat at constant volume, L is the latent heat, g is the gravitational constant and PdV is the energy expended in expansion of the air.

As the moist air rises, water condenses and latent heat is released. Initially, this heats the air, increasing the heat content, $C_v \Delta T$. However, the air is an open thermodynamic system and any increase in temperature will cause the air to expand. $C_v \Delta T$ decreases as the internal energy is converted to work to expand the volume against the pressure, $P\Delta V$. The air now has more buoyancy and rises faster through the atmosphere. If the temperature of the moist air at the surface is increased, the specific humidity must increase to maintain a constant RH. This means that more latent heat is released at lower altitudes by the ascent of a warmer air parcel compared to one at lower temperature and the same RH. The overall effect of this process is that as the surface temperature is increased, the specific humidity in the lower troposphere increases. However, at higher altitudes the specific humidity is reduced because of enhanced condensation at lower altitudes. This is discussed in more detail in Reference 49.

4.6 LWIR Emission to Space from the Atmosphere

The LWIR absorption and emission in the atmosphere is produced by a large number of overlapping molecular lines as illustrated above in Figure 3-7. In the lower troposphere these are broadened by molecular collisions so that they behave as a quasi-continuous black body. However, in the middle to upper troposphere, as the pressure and temperature decrease, the effect of pressure broadening is reduced, the H_2O concentration decreases significantly and the absorption and emission from the individual lines must be considered. These have a Lorentzian profile.[27] The LWIR emission from the wings of the broadened lines at lower altitudes is not reabsorbed by the narrower lines at higher altitudes. This results in a gradual transition with increasing altitude from LWIR exchange (absorption and emission) to a free photon flux that is emitted to space. Conversely, the downward emission from the narrower lines in the upper atmosphere must be absorbed and re-emitted by the broader molecular lines at lower altitudes. This means that the upward and downward LWIR fluxes are not equivalent. This is illustrated in Figure 4-48. This also invalidates the radiative forcing assumption of flux equilibrium.

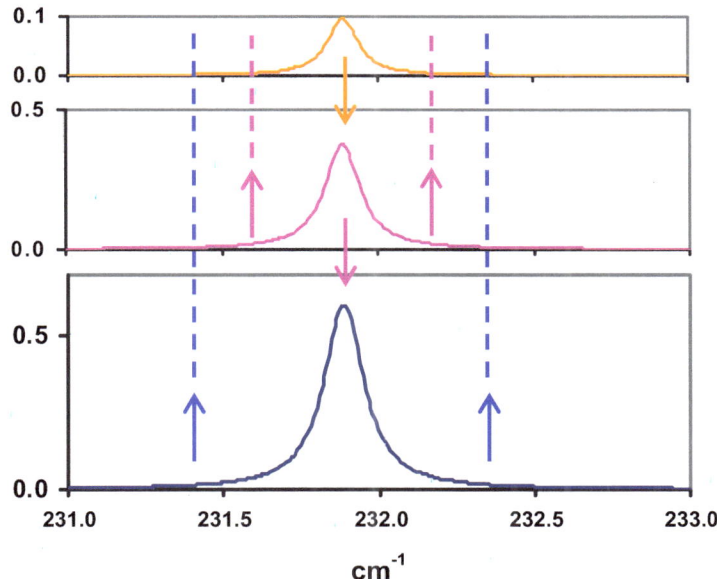

Figure 4-48: Change in linewidth with altitude. Photons emitted from the wings of the pressure broadened lines in the lower troposphere are not reabsorbed at higher altitudes.

The LWIR flux leaving the top of the atmosphere originates from a wide range of altitudes within the atmosphere as the gradual transition from LWIR exchange to a free photon flux occurs. Because of the differences in linewidth, pressure broadening, line spacing and concentration, the LWIR photon fluxes produced by H_2O and CO_2 have different altitude emission profiles. Figure 4-49 shows a plot of the calculated net emission vs. altitude up to 9 km for surface/air temperatures of 297/295 K, 287/285 K and 277/275 K with the RH set to 50 %. The plot shows net absorption and the negative sign means that there is net emission to space and therefore cooling. The altitude resolution is 100 m. The plot shows the net upward emission (cooling) for each 100 m layer. The approximate contributions of H_2O and CO_2 to the total emission are plotted separately. The H_2O contribution was estimated by adding the flux from the spectral regions 200 to 550 cm^{-1} and 1200 to 2000 cm^{-1}. The CO_2 contribution was estimated from the total flux in the spectral region from 550 to 800 cm^{-1}. This spectral region also includes some H_2O lines. There is a 'spike' near 1.7 km which marks the transition of the lapse rate from -6.5K.km^{-1} to the saturated lapse rate. This produces a minor discontinuity in the radiative transfer model output. The H_2O absorption lines show a peak cooling rate of ~-1.5 W.m^{-2}. This is the net upward free photon LWIR emission from a 100 m layer. The emission peak shifts to lower altitude as the surface/air temperatures decrease, but the emission profile is similar for all cases. This is because the peak emission comes from an emission band in the H_2O concentration range of 10^{16} to 10^{17} molecules.cm^{-3}. The corresponding temperatures are in the 240 to 260 K range. Since there is only a 2 K difference between the surface and air temperatures, the net emission to space starts close to the surface, above 100 m. There can be no water vapor 'feedback'. From 1 to 9 km, the total cumulative H_2O cooling rate was over 2.5 larger than that of CO_2 at all surface temperatures.

Figure 4-50 shows the absorbed flux vs. altitude for a surface/air temperature of 325/295 K and RH of 90, 50 and 10 %. The emission peak increases from an altitude of approximately 6 km to 7 km as the RH increases. The saturation altitude for the 10% RH case is at 5 km. There is a net absorption at altitudes below 1 km because of the higher surface temperature compared to Figure 4-49.

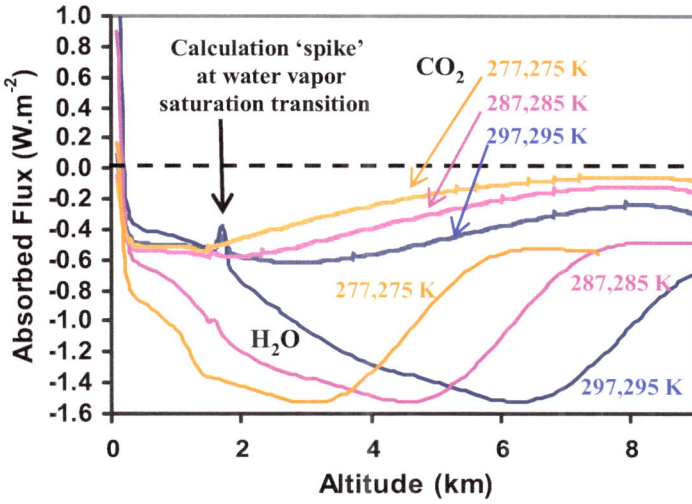

Figure 4-49: Absorbed flux vs. altitude for the surface/air temperatures of 297/295; 287/285 and 277/275 K, 50% RH and 100 m altitude resolution. The negative flux means that there is a net emission to space. The approximate contributions of H_2O and CO_2 to the total emission are shown separately (truncated data set).

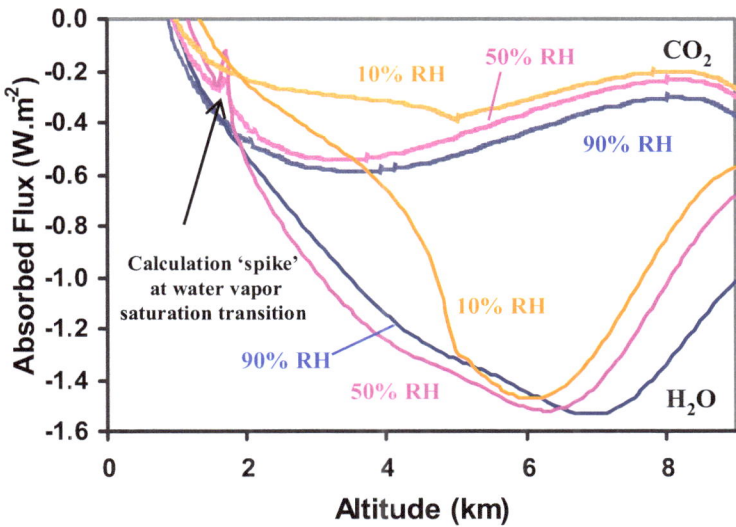

Figure 4-50: Absorbed flux vs. altitude for the surface/air temperature of 325/295 at 10, 50 and 90% RH and 100 m altitude resolution. The negative flux means that there is a net emission to space. The approximate contributions of H_2O and CO_2 to the total emission are shown separately (truncated data set).

Figure 4-51 shows the effect on the absorption-emission profile of increasing the CO_2 concentration from 200 to 280, 380 and 500 ppm for surface/air temperatures of 325/295K, 297/295 K, 287/285 K and 277/275 K with the RH set to 50 %. The peak emission occurs at altitudes between ~1 and 3 km. The magnitude of the peak emission is between 0.5 and 0.65 $W.m^{-2}$. An increase of 100 ppm from 280 to 380 pp in CO_2 concentration decreases the peak cooling flux by approximately 0.017 $W.m^{-2}$. For a 100 m column of air with a heat capacity of 1 $kJ.m^{-3}$, this produces a change in the cooling rate of ~0.0006 $C.hr^{-1}$ or 0.015 C per day. For a total cooling flux of 2 $W.m^{-2}$, the cooling rate for a 100 meter column of air is 0.07 $C.hr^{-1}$ or 1.7 C per day. A 100 ppm increase in atmospheric CO_2 concentration can have no measurable effect on the dynamic energy balance that determines the atmospheric temperature profile.[51]

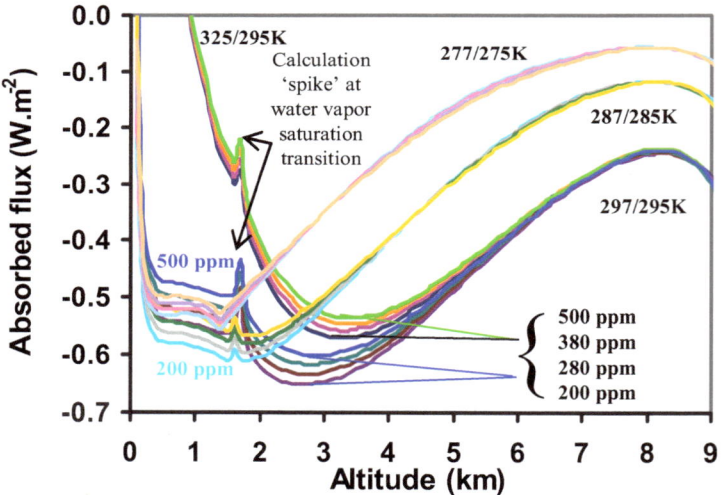

Figure 4-51: The effect of changes in CO_2 concentration on the absorbed flux vs. altitude for the surface/air temperatures of 325/295; 297/295; 287/285 and 277/275 K, 50% RH and 100 m altitude resolution. The negative flux means that there is a net emission to space. Estimated CO_2 emission in the 550 to 800 cm^{-1} range, (truncated data set).

5.0 THE EARTH'S DYNAMIC RADIATION BUDGET

The static, average energy balance diagrams of the Earth's radiation budget such as those published by the IPCC conceal the dynamic aspects of the energy transfer and imply a nonexistent equilibrium climate state.[52] The sun only heats the surface during the day. To illustrate the time dependent, dynamic nature of the energy transfer, the energy balance has to be separated into an average12 hour convective cycle and an average 24 hour LWIR emission cycle. This is shown in Figure 5-1. The 12 hour averages for the solar flux are just twice the static IPCC averages. The net average solar flux reaching the surface is 336 W.m^{-2} in 12 hours. This heats the surface and drives the daytime moist convection. The convection then heats the two atmospheric thermal reservoirs. The lower reservoir, the first 1 to 2 km layer of the troposphere provides almost all of the downward LWIR flux at the surface. It acts as a 'thermal blanket' and is not strongly coupled to the upper atmospheric thermal reservoir. It is the thermal storage in this layer that provides the 'greenhouse effect'. The surface air temperature of this layer also controls the end of the daytime convection cycle as discussed above in Section 4.1.3. This in turn is a dominant factor in setting the night time surface temperature. The upper reservoir is cooled by the LWIR emission to space. The emission to space is dominated by the molecular line narrowing of the water vapor emission. The water vapor concentration is controlled by the atmospheric temperature profile as illustrated above in Figure 4-49. This is reset each day by convection and atmospheric transport (weather). The 24 hour average direct LWIR window transmission flux to space stays the same as in the static energy balance. The primary purpose of this diagram is to change the way we think about climate energy transfer and show the underlying physics. The numbers just indicate the approximate long term dynamic energy balance. There is no thermal equilibrium on any time scale. The energy transfer is determined by the First and Second Laws of Thermodynamics. The First Law requires an energy balance. The Second Law requires a thermal gradient for heat transfer. When this type of diagram is used, it is clear that a doubling of the atmospheric CO_2 concentration can have no effect on the surface temperature. Furthermore, there can be no CO_2 'signature' in the MSAT record.

The dynamic fluxes for the six energy transfer processes have to be coupled into four thermal reservoirs: the ocean, the land surface, and the lower and upper atmospheric thermal reservoirs. This is illustrated in Figure 5-2. The long term changes in the daily and seasonal heat content of these thermal reservoirs determines the Earth's climate. Since the ocean thermal reservoir is approximately 2 orders of magnitude larger than the other reservoirs, it should be clear that changes in ocean thermal storage are the source of climate change. The dynamic approach also introduces another measurable climate variable. This is the phase shift or time delay between the peak of the solar flux and the peak temperature. This should also be predicted correctly by properly validated climate simulation models.

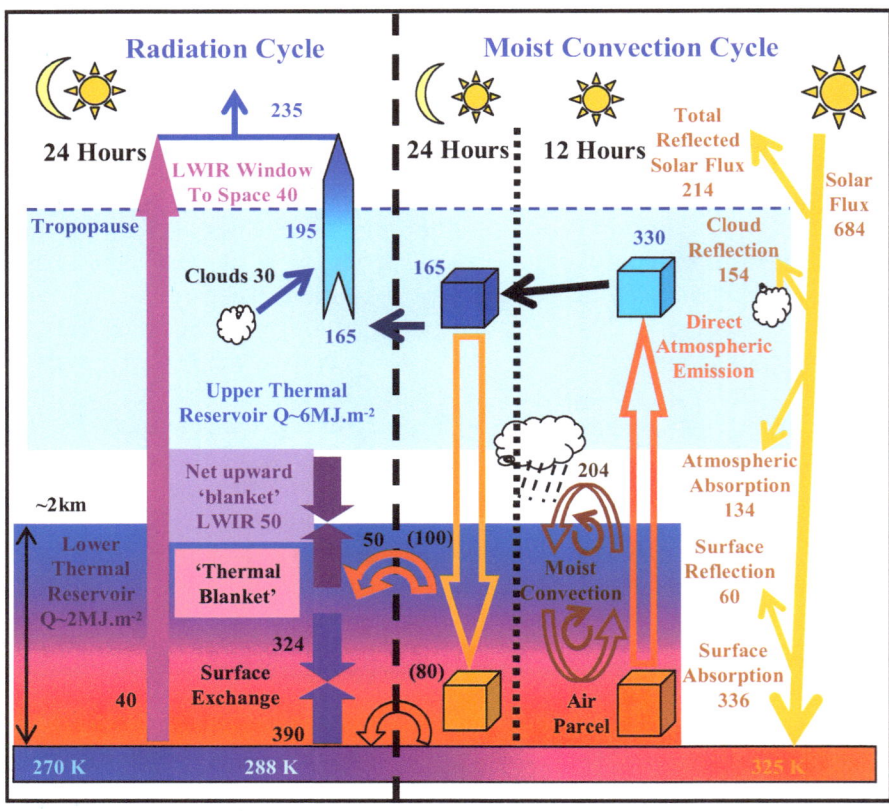

Figure 5-1: The Earth's radiation budget for the dynamic greenhouse effect. The surface energy is transported through the troposphere by moist convection shown as a 12 hour average cycle. LWIR emission to space occurs all the time and is shown as a 24 hour average cycle. The surface temperature is maintained by a thermal blanket about 2 km thick that is heated from the surface by convection and excess LWIR absorption. There is a dynamic energy balance between the two cycles. A doubling of the atmospheric CO_2 concentration can have no effect on the surface temperature.

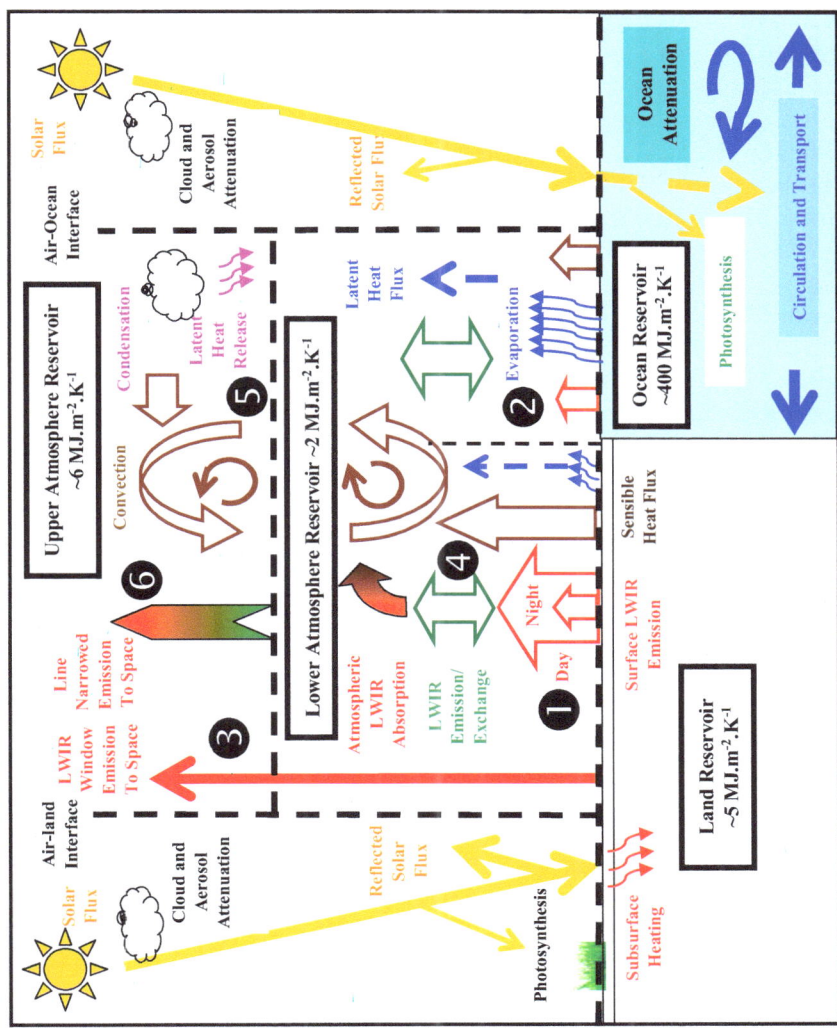

Figure 5-2: The energy transfer processes and thermal reservoirs required for the dynamic greenhouse effect (schematic).

The dynamic greenhouse effect also leads to a climate averaging paradox. All of the numerical analysis presented in this monograph used time steps from 1 minute for the land thermal conduction model to 1 hour for the diurnal ocean surface temperature calculations. The changes in ocean temperatures over 450 years were calculated using 30 minute time steps. Although climate trends are generally determined over a period of 30 years or longer, this trend is the result of the accumulation of a long period of short term changes. In order to produce realistic results, climate models must simulate the short term changes and then take the long term average of these short term changes. There is no 'shortcut' calculation of an equilibrium climate state that can be substituted for the real long term climate average.

The persistence of the cult of global warming comes from economic and political forces. The scientific hypothesis of radiative forcing was introduced in the mid 1960's as an honest attempt to explain climate change based on pre-satellite data and very limited computer capabilities.[15] It should have been rejected as invalid over 30 years ago as a part of the normal process of scientific advancement as new results become available. Instead, the MSAT record was substituted for the real surface temperature and the pseudoscience of the global warming cult was created.[4] The independence of the peer review process in climate science has been severely compromised. This has impeded progress in climate research for several decades. A new set of properly validated climate simulation models are needed that do not rely on the pseudoscience of radiative forcing. In order to incorporate the real physics of the surface energy transfer into these models, additional data are required. Meteorological measurements need to be upgraded to include mass and energy flux and the actual surface temperature. The Argo Float Program needs to be expanded to measure the coupling of the solar flux into the oceans. The dynamic energy transfer processes in the upper troposphere, particularly cloud dissipation also require further study. In addition, a better understanding of the variations of the solar flux itself is needed. Once the dynamic nature of climate energy transfer is properly understood, progress in climate science should be quite rapid. However, this requires a paradigm shift in the way we perceive climate change. The concepts of climate equilibrium and radiative forcing must be firmly rejected as invalid hypotheses. The 'greenhouse effect' is dynamic, not static. There is no climate equilibrium and no CO_2 induced global warming.

REFERENCES

1. Fourier, B. J. B., <u>Mem. R. Sci. Inst.</u> **7** 527-604 (1827), 'Memoire sur les temperatures du globe terrestre et des espaces planetaires' [Translation available at: http://www.wmconnolley.org.uk/sci/fourier_1827/]

2. Tyndall, J., Proc. Roy Inst. Jan 23 pp 200-206 (1863), 'On radiation through the Earth's atmosphere'

3. Clark, R., <u>Energy and Environment</u> **21**(4) 171-200 (2010), 'A null hypothesis for CO_2'

4. Hansen, J. et al, (45 authors), <u>J. Geophys. Research,</u> **110** D18104 pp1-45 (2005) 'Efficacy of climate forcings' http://pubs.giss.nasa.gov/authors/jhansen.html

5. Alley R. B., et al, (51 authors), *IPCC, Summary for Policymakers, Climate Change 2007: The Physical Science Basis.* Contribution of Working Group I to the Fourth Assessment, eds: Solomon, S., Qin, D., Manning, M., Chen, Z., Marquis, M., Averyt, K. B. Tignor, M. and Miller, H. L., *Report of the Intergovernmental Panel on Climate Change*, Cambridge University Press, Cambridge, United Kingdom and New York, NY, USA., 2007.

6. Cheetham, A. 'A history of the global warming scare', http://scienceandpublicpolicy.org/reprint/history_of_global_warming_scare.html

7. F. W. Taylor, *Elementary Climate Physics*, Oxford University Press, Oxford, 2006. [Chapter 7]

8. Agassiz, L. *Etudes sur les Glaciers*, Neuchatel, Paris, (1840).

9. Imbrie, J. and Imbrie, K. P., '*Ice Ages, solving the mystery*', Harvard University Press, Cambridge, Mass. 1979.

10. Zachos, J.; M. Pagani, L. Sloan, E. Thomas and K. Billups, <u>Science</u> **292** 686-693 (2001), 'Trends, rhythms and aberrations in the global climate 65 Ma to present'

11. Arrhenius, S. <u>Philosophical Transactions</u> **41** 237-276 (1896) 'On the influence of carbonic acid in the air upon the temperature of the ground'

12. Callendar, G. S., <u>J. Royal Met. Soc.</u> **64** 223-240 (1938), 'The artificial production of carbon dioxide and its influence on temperature'

13. Revelle, R. and H. E. Suess, <u>Tellus</u> **9**: 18-27 (1957), 'Carbon dioxide exchange between atmosphere and ocean and the question of an increase of atmospheric CO_2 during the past decades'

14. http://scrippsco2.ucsd.edu/data/in_situ_co2/monthly_mlo.csv. Keeling curve data.
15. Manabe, S. and R. T. Wetherald, J. Atmos. Sci. **24** 241-249 (1967), 'Thermal equilibrium of the atmosphere with a given distribution of relative humidity'
16. Weart, S. R., Physics Today **50**(1) 34-40 (Jan 1997), 'The discovery of the risk of global warming'
17. Ramanathan, V., Ambio **27**(3) 187-197 (1998), 'Trace gas greenhouse effect and global warming'
18. Jones, P. D., New, M., Parker, D. E., Martin, S. and Rigor, I. G., Rev. Geophysics, **37**(2) 173-199 (1999), 'Surface air temperature and its changes over the past 150 years'
19. http://data.giss.nasa.gov/gistemp/graphs Contiguous 48 U.S. Surface Air Temperature Anomaly.
20. D'Aleo, 'J Effects of AMO and PDO on temperatures' Intellicast, May 2008. http://www.intellicast.com/Community/Content.aspx?a=127
21. D'Aleo, J. 'Progressive Enhancement of Global Temperature Trends', Science and Public Policy Institute, July 2010. http://scienceandpublicpolicy.org/originals/progressive_enhancement.html
22. W. Ferrell, Nashville J. Medicine and Surgery **Vol xi**, No. 4 and 5, Oct. Nov (1856), 'An essay on the winds and currents of the ocean'
23. C. G. Rossby, Quarterly Journal of the Royal Meteorological Society **66** 68-87 (1940), 'Planetary flow patterns in the atmosphere'
24. W. Munk, and D. Day, Oceanography **15**(4) 7-29 (2002), 'Harald Sverdrup and the War Years'
25. D'Aleo, J. and D. Easterbrook, 'Multidecadal tendencies in Enso and Global Temperatures Related to Multidecadal Oscillations' http://scienceandpublicpolicy.org/reprint/multidecadal_tendencies.html
26. http://rredc.nrel.gov/solar/spectra/am1.5/ASTMG173/ASTMG173.xls ASTM Reference Solar Spectra
27. Rothman, L. S. et al, (30 authors), J. Quant. Spectrosc. Rad. Trans. **96** 139-205, (2005), 'The HITRAN 2004 molecular spectroscopic database'
28. NASA, *U. S. Standard Atmosphere*, NASA-TM-X-74335, 1976
29. A. A. Tsonis, *An Introduction to Atmospheric Thermodynamics*, 2nd edn., Cambridge University Press, Cambridge, UK, 2007. [p. 127.]
30. Jenkins, F. A. and H. E. White, *Fundamentals of Optics*, McGraw Hill, NewYork, NY, 1976, 4th Ed. [Chapter 25]
31. Hale, G. M. and M. R. Querry, Applied Optics **12** 555-563 (1973), 'Optical constants of water in the 200 nm to 200 μm region'

32. Fisk, C., http://www.climatestations.com/los-angeles/, 'Graphical Climatology of Downtown Los Angeles: Daily Temperatures and Rainfall, by Year (1921 - Present)' [The numerical data for the analysis presented here was kindly provided by the author.]
33. Billo, E. J., *Excel for Scientists and Engineers*, Wiley Interscience, Hoboken, NJ, 2007 pp269-273.
34. Baldocchi, D. D., Global Change Biology **9** 1-14 (2003), 'Assessing the eddy covariance technique for evaluating carbon dioxide exchange rates of ecosystems: past, present and future'
35. Clark. R., 'What surface temperature is your model really predicting?' http://hidethedecline.eu/pages/posts/what-surface-temperature-is-your-model-really-predicting-190.php
36. Clark, R., 'California climate change is caused by the Pacific Decadal Oscillation, not by carbon dioxide' http://scienceandpublicpolicy.org/originals/pacific_decadal.html
37. http://public.ornl.gov/ameriflux/index.html, AmeriFlux Network
38. http://jisao.washington.edu/pdo/PDO.latest, Pacific Decadal Index from 1900.
39. http://www.wrcc.dri.edu/Climsum.html, Western Region climate data.
40. http://www.piercecollegeweather.com/, Pierce College weather station data.
41. http://www.metoffice.gov.uk/climate/uk/stationdata/, UK Historical Climate Data
42. http://www.cru.uea.ac.uk/cru/data/temperature/, HadSST2 Sea Surface Temperature Database.
43. Yu, L., Jin, X. and Weller R. A., *OAFlux Project Technical Report* (OA-2008-01) Jan 2008, 'Multidecade Global Flux Datasets from the Objectively Analyzed Air-sea Fluxes (OAFlux) Project: Latent and Sensible Heat Fluxes, Ocean Evaporation, and Related Surface Meteorological Variables' (Available at: http://oaflux.whoi.edu/publications.html)
44. http://oaflux.whoi.edu/images2_flux/EV_50a.jpg
45. http://floats.pmel.noaa.gov/index.html, *Argo Profiling CTD Floats*, NOAA Pacific Marine Environmental Laboratory.
46. Levitus, S., J. Antonov and T. Bower, Geophysical Research Letters , **32** L02604 1-4 (2005), 'Warming of the world ocean 1955-2003'
47. Yu, L., J. Climate **20**(21) 5376-5390 (2007), 'Global variations in oceanic evaporation (1958-2005): The role of the changing wind speed' (Available at: http://oaflux.whoi.edu/publications.html)
48. Seidel, D. J.; M. Free and J. Wang, J. Geophys Res. **110** D090102 1-13 (2005), 'Diurnal cycle of upper air temperature estimated from radiosondes'

49. Gilbert, W.C., <u>Energy and Environment</u> **21**(4) 263-276 (2010) 'The thermodynamic relationship between surface temperature and water vapor concentration in the troposphere'

50. Lindzen, R.S. and Y-S. Choi, <u>Geophys Res. Letts</u> **36** L16705 1–6 (2009) 'On the determination of climate feedbacks from ERBE data'

51. Clark. R., Gravity rules over the photons in the greenhouse effect. <u>http://hidethedecline.eu/pages/posts/greenhouse-effect-vs.-gravity---guest-post-by-roy-clark-201.php</u>

52. Kiehl, J.T. and K. E. Trenberth, <u>Bull. Amer. Meteor. Soc.,</u> **78**(2) 197-208 (1979), 'Earth's annual global mean energy budget' (FAQ 1.1, Figure 1, p. 96 IPCC Fourth Assessment Report, 2007)

About the Author

Dr. Roy Clark is President and Founder of Ventura Photonics. He has over 30 years of experience in optics and spectroscopy with emphasis on new product and process development for adverse environments. His experience includes optical and spectroscopic sensors, combustion and laser diagnostics and non imaging optics for illumination and solar concentrators. He holds 8 US patents.

This book is the result of research that the author began in 2007, following the publication of the 2007 IPCC Fourth Assessment Report. The IPCC Report did not provide a quantitative explanation of carbon dioxide induced climate change. All of the arguments were empirical and circular. Carbon dioxide must cause global warming, therefore more carbon dioxide must cause more warming. The climate models were hard wired to produce the desired results. The question that the author asked was a simple one. How does a 100 ppm increase in the atmospheric concentration of carbon dioxide cause a 1 C rise in the measured meteorological surface air temperature? The answer that he found from his research is also simple. It is impossible for the resulting 1.7 $W.m^{-2}$ increase in the downward LWIR flux from carbon dioxide to cause any climate change. This follows from a quantitative analysis of the dynamic surface energy transfer that is the topic of this monograph.

In 1827, Fourier clearly understood the dynamic nature of the surface energy transfer. This work provides quantitative confirmation of the dynamic heating effects that he described almost 200 years ago.

www.ingramcontent.com/pod-product-compliance
Lightning Source LLC
Chambersburg PA
CBHW040808200526
45159CB00022B/53